(continued on inside back cover)

THE HANDBOOK

OF TECHNICAL

WRITING

Form and Style

THE HANDBOOK OF TECHNICAL WRITING

Form and Style

- Mary Lee
- Gloria Stephenson
- Max Anderson
- Lynn Allan Lee

University of Wisconsin, Platteville

Harcourt Brace Jovanovich, Publishers

San Diego   New York   Chicago   Austin
Washington, D.C.   London   Sydney
Tokyo   Toronto

PREFACE

The Handbook of Technical Writing: Form and Style is a comprehensive yet concise reference guide for technical writing. Students in technical writing courses and people already working in business, industry, and government can use the *Handbook* as a ready resource for information about all aspects of technical communication as well as the essentials of conventional English. In the classroom, instructors may use the *Handbook* with a standard textbook, but the *Handbook* can easily stand alone as the major textbook.

We feel that the *Handbook*, as a practical reference guide in technical communication, is unique in its clarity of organization, which makes the information easily accessible not only to those who are familiar with the terminology but also to those, such as students, who are not. We use a topical arrangement of material, so readers can find the solution to a problem without having to know a specific term. The reader can locate the needed section by using the handy guide on the inside of the cover, the table of contents, or the index, as well as the margin tabs that denote section numbers. Each topical division enumerates all the sections under that topic. If, on the other hand, the reader already knows a term, the guide or index can lead directly to the specific entry.

The first half of the *Handbook* stresses the essentials of grammar, punctuation, mechanics, diction, and sentence structure, but moves beyond standard English grammar texts by including material specific to scientific and technical writing. For example, the part titled Mechanics demonstrates preferred usage for writing numbers, abbreviations, measurements, symbols, and equations. The discussions on diction and sentence structure pay particular attention to the style of technical documents. The examples throughout the book are taken from business, industry, science, and technology.

The second half of the book contains many types of technical communication, and division topics such as Documentation and Graphics concentrate on relating these subjects to scientific and technical writing. In many cases, techniques and strategies are illustrated with documents actually produced in business situations.

Throughout the *Handbook* we seek to describe the forms, style, and usage for the most effective technical writing. In our efforts to produce a thorough and accurate book, we have received indispensable support from our editors at Harcourt Brace Jovanovich: Stuart Miller, Kay Kaylor, and Brett Smith. We also thank the designer, Gina Sample; art editor, Cindy Robinson; and production manager, Lesley Lenox. We owe a special debt of gratitude to our first editor at HBJ, Paul H. Nockleby, with whom Mary Lee originally shared her vision for the book. We wish to note further the sustaining support of Rosemary Anderson. And we owe a very special thanks and a debt of gratitude to Roger Stephenson for his important technical contributions and advice.

Mary Lee

Gloria Stephenson

Max Anderson

Lynn Allan Lee

CONTENTS

THE HANDBOOK
OF TECHNICAL
WRITING
Form and Style

GRAMMAR

●1

NOUNS

KEY POINTS

- Distinguish correctly between plural and singular collective nouns and between common and proper nouns
- Do not overload your writing with nouns. This will only test your reader's patience.
- Do not rely on nouns to convey the actions that verbs convey.
- Use nouns selectively for greatest impact.

A noun is the part of speech that names a person, place, object, or concept. A noun may be either common or proper.

EXAMPLES

Common Nouns	Proper Nouns
company	General Motors
area	New England
university	Harvard University
holiday	Christmas
software	Lotus 1–2–3
tool	Phillips screwdriver

Because much technical writing deals with specific companies and products, you should be aware of several problem areas in noun usage.

1.1   LEARN TO RECOGNIZE COLLECTIVE NOUNS.

A collective noun refers to a group. Treat the noun as singular when it refers to the group as a whole. When a collective noun refers to members of a group acting individually, treat the noun as plural.

EXAMPLES

Singular

The jury is agreed on a verdict. (All the members agree.)
The jury is staying at the hotel. (All the members are staying at the same hotel.)

Plural

The jury are at loggerheads. (The members of the jury do not agree.)
The jury are going home. (Each jury member is going to his or her own home.)

Note: In a few cases, you have the option of using either singular or plural verbs and/or pronouns.

EXAMPLES

The faculty *have/has* demanded a pay raise.
The crew *is/are* working on *its/their* project one more week.
One hundred dollars *is/are* lying on the table.

1.2   AVOID NOUN-HEAVY PHRASES UNLESS YOU HAVE A VERY GOOD REASON TO USE ONE.

A noun phrase overuses nouns by grouping several together.

EXAMPLES

The New York sheet metal contractor contestant won the lottery.
The company purchased the plasma arc cutting table-part development system.

Note: At times noun-heavy phrases may be nearly indecipherable and should be improved.

1.5

Weak

graduate education program analysis system
executive management potential performance feasibility study

Improved

graduate catalogue
executive feasibility study

1.3 DO NOT CONVERT NOUNS TO VERBS.

Using trade names as verbs has become very popular in advertising and other businesses. In informal speech this practice may be acceptable, but in technical documents it is improper usage. Use standard verb forms to express actions.

Weak	Improved
I will Xerox the document for you.	I will photocopy the document for you.
He Simonized his car.	He waxed his car.

1.4 DO NOT USE TRADE NAMES AS COMMON NOUNS.

Although some trade names have become synonymous with the product, do not use them to replace a generic word that may refer to any trade name.

Weak Use of Trade Name	Improved Use of Generic Term
I need a Kleenex.	I need a facial tissue.
Jesse bought a new pair of Levis.	Jesse bought a new pair of jeans.
Gene bought some Scotch tape.	Gene bought some transparent adhesive tape.

1.5 DO NOT RELY ON NOUNS TO CONVEY ACTION.

Verbs are much better suited for this purpose.

Weak	Improved
Many college roommates are in a social agreement that allows them to coexist in the same room.	Many college roommates agree to coexist in the same room.

Busy executives place themselves in need of relief from the endless routine of report writing.	Busy executives must break away from the endless routine of report writing.

●**EXERCISE 1**   In each of the following sentences, correct any faulty noun usage.

1. Many college graduates situate themselves in positions that allows for little job satisfaction.
2. It is time to call Ma Bell about the phone service.
3. The family of humpback whales are protected around the world.
4. The manager was proud of the maximum fixed performance potential estimates guide.
5. The board vote every Tuesday night.
6. Bring me a glass of Minute Maid.

●2

VERBS

- Whenever possible, use active rather than passive verbs.
- Use the correct verb tense.
- Avoid relying too heavily on forms of *to be*.
- Learn to correctly use *lie/lay* and *sit/set*.
- Identify the subject and verb in your sentences to check for correct agreement.
- Use standard verbs, not verbs coined from other parts of speech.

A verb is the part of speech that suggests an action or something taking place or existing in the past, present, or future. Since verbs, more than any other part of speech, make writing interesting and clear, it is important to be aware of the different verb forms. Know the varied and correct uses of verbs, both in tense and agreement and in coinage from other parts of speech.

A technical writer struggling with verb forms might wonder why so many different tenses are necessary. Precision writing of sequences of operations often can be described only with careful use of sequences of verb tenses. If a writer uses verbs skillfully, the writing will improve in both clarity and interest.

2.1   USE FINITE OR NONFINITE VERB FORMS.

a. Use a finite verb as the main verb in a sentence.

A finite verb is either *transitive* or *intransitive*. A *transitive verb* takes a direct object and is in the active voice; that is, the subject acts rather than being acted upon. An *intransitive verb* does not take a direct object and is frequently, but not always, in the passive voice; that is, the subject is acted upon.

EXAMPLES

Transitive Verbs	Intransitive Verbs
The engineer read the blueprints. [active]	The blueprints were read by the engineer. [passive]
The men dug a ditch. [active]	The ditch was dug by the men. [passive]
The workers installed a new roof. [active]	The roof was installed by the workers. [passive]
The new owner cut the workers' salaries. [active]	The workers' salaries were cut by the new owner. [passive]

Note: In general, transitive verbs are more forceful than intransitive verbs. "Linking" verbs, such as *was, seems, looks,* and *smells,* connect subject and complement and are always intransitive.

b. Use nonfinite verbs as nouns, adjectives, or adverbs.

A *nonfinite verb,* commonly called a *verbal,* is a word derived from a verb that functions as a noun, adjective, or adverb. A verbal cannot function as the main verb of a sentence. The three types of verbals are called *gerunds, infinitives,* and *participles.* Distinguish them from one another as follows:

Gerunds   Gerunds are verbals ending in *-ing* that function as nouns. Use them as subjects or objects in a sentence.

EXAMPLES

Smoking endangers your health. [subject]
He disappeared by *moving* quickly. [object of a preposition]

Scientists maneuvered the *drifting* satellite back on course. [direct object]

Infinitives   Infinitives are verbs preceded by *to* that may function as nouns, adjectives, or adverbs.

EXAMPLES

Sally loves *to program* word games. [infinitive as noun]
Soldier ants fight *to win*. [infinitive as adverb]
Automotive engineering is the career *to choose*. [infinitive as adjective]

Participles   Participles are verbals ending in *-ed, -en, -nt, -rn,* or *-t* in the past tense or *-ing* in the present tense. They function as adjectives. Unlike gerunds, participles cannot be the subject of a sentence, nor can they be used as an object.

EXAMPLES

The *striking* labor force is angry. [present participle]
The *overused* copy machine broke down. [past participle]

Note: Finite verbs are usually stronger than infinite verbs, but the skillful use of infinite verbs avoids awkward constructions and makes ideas clearer.

Awkward	**Improved**
The bond issue was voted on by the town, which turned against its beliefs.	*Voting* for the bond issue, the town turned against its beliefs.

●**EXERCISE 1**   Identify the verb forms in the following sentences as finite or infinite. If a verbal is used, identify it as a gerund, infinitive, or participle.

1. The batteries run down in my lap computer.
2. John fixed the broken printer.
3. The space lab stopped tumbling after three hours.
4. The Namib desert is filled with living creatures.
5. The bags were carried by the junior-level engineer.
6. The basatic disk, weighing 24 metric tons, was discovered in 1790.

7. Writing a progress report can be time consuming.
8. That beetle is called little *cardiosis.*
9. Frank likes to run every morning.

2.2 EMPLOY THE CORRECT VERB TENSE.

Verb tense is the most confusing aspect of verb usage. Always remember that the three basic tenses are present, past, and future. Each indicates an action taking place at a specific time period.

For the most part, the principal parts of a verb, determined by tense, follow a regular pattern.

Simple Tenses	Perfect Tenses
present: bake (bakes)	present: have (has) baked
past: baked	past: had baked
future: will (shall) bake	future: will (shall) have baked

It is not enough to know the verb tenses; it is also necessary to use them correctly in complete sentences.

EXAMPLES

Present

The engineer *asks* the contractor many questions.

Past

The engineer *asked* the contractor many questions.

Future

The engineer *will ask* the contractor many questions.

Note: Present- and past-tense verbs use single-word verbs. Future-tense verbs require two words.

Three perfect tenses may be used to indicate a more precise time in the present, past, or future. Use the *present perfect* for a period prior to a designated time in the present. Use *past perfect* to indicate action completed prior to a designated moment in the past. Use the *future perfect* to indicate action completed before an indicated time in the future.

EXAMPLES

Present Perfect

> The engineer *has asked* the contractor many questions about the strength of the foundation.

Past Perfect

> The engineer *had asked* the contractor many questions about the strength of the foundation before it caved in.

Future Perfect

> The engineer *will ask* the contractor many questions about the strength of the foundation before it is poured.

Note: In the above sentences, the time when the engineer asked the contractor questions helps to clarify the engineer's position both before and after the mishap.

A number of verbs take irregular parts and are often misused. The most irregular verb in English is *to be*. The following chart gives the forms for *to be:*

	Simple Tenses	**Perfect Tenses**
present	I am	I have been
	you are	you have been
	he/she/it is	he/she/it has been
	we are	we have been
	you are	you have been
	they are	they have been
past	I was	I had been
	you were	you had been
	he/she/it was	he/she/it had been
	we were	we had been
	you were	you had been
	they were	they had been
future	I will be	I will have been
	you will be	you will have been
	he/she/it will be	he/she/it will have been
	we will be	we will have been
	you will be	you will have been
	they will be	they will have been

The following list gives the principal parts of a number of irregular verbs more often misused:

Present Tense	Past Tense	Past Participle
arise	arose	arisen
become	became	become
begin	began	begun
bet	bet	bet
bid (command)	bade	bidden
bind	bound	bound
blow	blew	blown
break	broke	broken
bring	brought	brought
burn	burned or burnt	burned or burnt
burst	burst	burst
buy	bought	bought
catch	caught	caught
choose	chose	chosen
cling	clung	clung
come	came	come
cost	cost	cost
cut	cut	cut
deal	dealt	dealt
dive	dived or dove	dived
do	did	done
drag	dragged	dragged
draw	drew	drawn
drink	drank	drunk or drank
drive	drove	driven
eat	ate	eaten
fall	fell	fallen
find	found	found
fly	flew	flown
forget	forgot	forgotten or forgot
freeze	froze	frozen
get	got	got or gotten
give	gave	given
gone	went	gone
grow	grew	grown
hang (an object)	hung	hung
keep	kept	kept

2.2

Present Tense	Past Tense	Past Participle
know	knew	known
lay (to put down)	laid	laid
lie (to recline)	lay	lain
lose	lost	lost
prove	proved	proved, proven
ride	rode	ridden
rise	rose	risen
run	ran	run
say	said	said
see	saw	seen
set	set	set
shake	shook	shaken
show	showed	shown or showed
shrink	shrank or shrunk	shrunk or shrunken
sink	sank or sunk	sunk
sit	sat	sat
speak	spoke	spoken
spend	spent	spent
spin	spun	spun
steal	stole	stolen
stink	stank or stunk	stunk
strive	strove	striven
swear	swore	sworn
take	took	taken
tear	tore	torn
think	thought	thought
throw	threw	thrown
understand	understood	understood
wake	woke or waked	waked or woken
wear	wore	worn
wind	wound	wound
wring	wrung	wrung
write	wrote	written

●EXERCISE 2 In each of the following sentences, choose the correct verb.

1. When the wildlife refuge *opened/had opened*, thousands of migrating birds stopped on their way south.

2. *Having sold/Selling* his new component, my boss was relaxed.
3. The company plans *to merge/to have merged* by next week.
4. The students arrived before the class *started/had started*.
5. America's last Venus probe *encountered/had encountered* enormous searing heat and crushing atmospheric pressure.
6. In 1944, O. T. Avery and colleagues *discovered/had discovered* DNA.

2.3   USE *SIT/SET* AND *LAY/LIE* CORRECTLY.

Probably more mistakes are made with these verbs than with any others. Each one refers to similar actions, but the differences are clear if a person knows the correct forms for each verb.

a. Be aware of the distinction between *sit* and *set*.

Sit means "to be seated" while set means "to put or place something down." *Set* needs a direct object while sit does not. Their verb forms follow:

Present	Past	Present Participle	Past Participle
set	set	setting	set
sit	sat	sitting	sat

EXAMPLES

He *sets* the book on the table.
He *set* the book on the table.
He is *setting* the book on the table.
He had *set* the book on the table.

He *sits* in my chair.
He *sat* in my chair.
He is *sitting* in my chair.
He had *sat* in my chair.

b. Know the distinction between *lay* and *lie*.

Lay means "to put something down" while *lie* means "to rest or to get into a horizontal position." *Lay* needs a direct object while *lie* does not.

Present	Past	Present Participle	Past Participle
lay	laid	laying	laid
lie	lay	lying	lain

EXAMPLES

He *lays* the book on the table.
He *laid* the book on the table.
He was *laying* the book on the table.
He had *laid* the book on the bed.

He *lies* down on the bed.
He *lay* down on the bed.
He was *lying* on the bed.
He had *lain* on the bed.

● **EXERCISE 3** In each of the following sentences, choose the correct form of *sit/set* or *lay/lie*.

1. He *sat/set* the transit in the truck.
2. I will *lie/lay* down for a nap.
3. Let me *lay/lie* the report on Paul's desk.
4. I have been *sitting/setting* at the computer all day.
5. The lazer had *laid/lain* there for three hours.
6. What is that object *laying/lying* on the ground?
7. He *sat/set* at the head of the table.
8. Who has been *laying/lying* on my bed?
9. *Sit/Set* the mail sack in the corner.
10. The new keyboard *sat/set* in that cabinet for ten days.

2.4 BE SURE SUBJECTS AND VERBS AGREE IN A SENTENCE.

Faulty subject-verb agreement is one of the most common problems in verb usage. It usually occurs because the writer is not certain what the subject of the sentence is.

a. Use a singular verb with a singular subject and a plural verb with a plural subject.

EXAMPLES

The *crane drops* the boxes on the floor. [singular subject and verb]
The *cranes drop* the boxes on the floor. [plural subject and verb]
The *programmer returns* the disks. [singular subject and verb]
The *programmers return* the disks. [plural subject and verb]

Note: All third-person-singular verbs take an *-s* ending, but *I*, though singular, does not.

b. Use the correct verb when a word or words come between the subject and verb.

One major reason for subject-verb disagreement is the user's inability to recognize the subject because a word or words may intervene between the subject and verb.

EXAMPLES

The *meeting* of the various companies *helps* to clarify the issues.
    [subject]   [prepositional phrase]   [verb]

The *administrators* who represent the owners *run* the company.
    [subject]   [dependent clause]   [verb]

c. Use *each, either, neither, one, everybody, nobody, none, no one,* and *anyone* with singular verbs even if one of these subjects obviously refers to more than one person.

EXAMPLES

Everybody is happy with the new contract.
Each of the directors *wants* to head the company.
Neither of the brothers *was* happy.
Anyone in the shipping department *is* a suspect.

d. Beware of subject-verb agreement with inverted word order; with the expletive *there*, verbs precede the subject.

EXAMPLES

At the end of the line *stood* three *engineers.*
            [verb]   [subject]

Last to speak *was* my *boss.*
       [verb]   [subject]

Among the greatest mistakes *is*   the *failure* to tolerate error.
             [verb]   [subject]

There *are* many *ways* to solve the problem.
    [verb]   [subject]

There *is* new *hope* for the future.
    [verb]   [subject]

●**EXERCISE 4**   In each of the following sentences, find the subject and then choose the correct verb.

1. The issue of importance *is/are* peace.
2. Everybody *is/are* happy with the new spreadsheet.

3. General Smith's commands *order/orders* the army into battle.
4. Either Blake or Wilson *was/were* to study the attack of the Orca whales.
5. There *is/are* hundreds of unmanned satellites that circle the earth.
6. None of the surveyors *is/are* overpaid.
7. The permanent records *go/goes* on to disks.
8. The corporation *is/are* ready to declare a dividend.
9. Of the three candidates for the position, he *is/are* preferable.

2.5 USE THE CORRECT MOOD TO EXPRESS THE ATTITUDE THE READER SHOULD HAVE TOWARD A SUBJECT.

Three moods are used in writing: the indicative, the imperative, and the subjunctive. The choice of mood will depend on whether the writer sees an assertion as a statement or question, a command or request, or a supposition or recommendation.

Indicative Mood   The indicative mood, the most common, states a fact or asks a question.

EXAMPLES

Barney is not afraid to work at night.
Where does Barney work?
Barney would like to resign.

Imperative Mood   The imperative mood states a command or makes a request. It is usually used with an understood subject.

EXAMPLES

Be quiet!
Work tonight.
Find the treasure on the island.

Subjunctive Mood   The subjunctive mood is used less frequently than in the past. It states a condition contrary to fact or makes a recommendation.

EXAMPLES

> If I were wealthy, I would invest in warehouses.
> I recommend that he invest in warehouses.
> Eva spoke as though she were the last member of her department.

Note: See Section 25, p. 160, for a more thorough discussion of sentence mood.

●**EXERCISE 5** Identify the mood for each of the following sentences.

1. Jump in the car.
2. Bring in the newspaper.
3. Jeff gave two reasons for volunteering for overtime.
4. If I were more ambitious, I would start working on the Benson report today.
5. Where are the bids for the Peterson job?
6. Carry the mail to the last desk on the right.
7. The citizens of the new world recommended independence.
8. Some of the people in the room were hostile.
9. Throw the ball!
10. Is there any reason to come back?

2.6 DO NOT OVERUSE *-IZE* ON THE END OF VERBS.

Some verbs end in *-ize*, for example, *realize, naturalize*. However, there is a growing tendency to create non-verbs by using *-ize*. This problem occurs frequently when nouns are used—very awkwardly—as verbs.

EXAMPLES

> Franklin will be *funeralized* (buried) on Monday.
> Let us *finalize* (finish) our contract.
> Our tonic will *energize* (improve) your health.

Note: These new "verbs" are unnecessary. They may look and sound impressive, but they demonstrate an ignorance of correct and clear verb usage.

●3

SENTENCES

KEY POINTS

- Use all four sentence types—simple, compound, complex, and compound-complex, but rely most on simple and complex sentences.
- Use the sentence type that best suits the purpose of your idea.
- When using subordination, be sure to subordinate the clause you do not want to emphasize.
- Use correct punctuation to avoid fragments, splices, and fused sentences.

Since the sentence is the basic unit of writing, you need to understand the basic sentence patterns. There are four types of sentences: *simple, compound, complex,* and *compound-complex.* Used with variety, they make your writing more interesting and more precise by allowing you to choose your tone and to emphasize the most important part of the sentence. (See Section 24 through Section 30, pp. 156–82, for more information on writing effective sentences.)

In order to understand sentence patterns, you first must know the two types of clauses that sentences are composed of: main and subordinate clauses. A *main clause* is a complete thought that contains a subject and a verb; the clause can stand by itself. For example, "I ran home" is a main clause. A *subordinate clause* is an incomplete thought that contains a subject and a verb but cannot stand by itself as a sentence. For example, "before I ran home" is a subordinate clause. Anyone reading "before I ran home" knows that something must have happened because the writer ran home, but the reader also needs a main clause that clarifies what caused the writer to run home.

18

Each of the four sentence types is identified by the number of and type of clauses in it.

The Simple Sentence   A simple sentence has one main clause.

EXAMPLES

The corporation declared a profit.
Lawyers generally like to plea bargain.

The Compound Sentence   A compound sentence has at least two main clauses.

EXAMPLES

The corporation declared a profit, and the stockholders were elated.
Lawyers generally like to plea bargain, but the public does not always favor this process and may become angry.

The Complex Sentence   A complex sentence has one main clause and at least one subordinate clause.

EXAMPLES

After the corporation declared a profit, the shareholders were elated.
Lawyers generally like to plea bargain since judges are inclined to be reasonable in such cases.

The Compound-Complex Sentence   A compound-complex sentence has at least two main clauses connected by a coordinating conjunction (*and, but, or, yet, so, for*) and at least one subordinate clause.

EXAMPLES

After the corporation declared a profit, the stockholders were elated, and the corporate executive officer received a large salary increase.
Lawyers generally like to plea bargain since judges are inclined to be reasonable in such cases, but the public does not always favor this process and often becomes angry.

Note: When using subordination, be sure to subordinate the right clause for proper emphasis.

Weak Emphasis	**Improved Emphasis**
When it calls for data by name, a program needs it.	When a program needs data, it calls for it by name.

3.1

●**EXERCISE 1** Label each of the following sentences as simple, compound, complex, or compound-complex.

1. The sales rep called, and I answered.
2. When the sales rep called, I answered.
3. When the sales rep called, I answered, and he asked if he could see me tomorrow.
4. The sales rep called.
5. Because of the war, gasoline was rationed.
6. The microchip revolutionized computers because it allowed them to be smaller, less expensive, and more powerful than earlier models.
7. The human skeleton has 206 bones.
8. Weather satellites help predict weather and track hurricanes, and communication satellites allow people all over the world to view events on television.
9. When the department bowling league was formed in 1960, ten teams entered, and in 1986 five of those remained.

Failure to understand main and subordinate clauses can lead to sentence errors: *fragments, comma splices,* and *fused* (run-on) *sentences*. Besides showing an ignorance of correct sentence structure, sentence errors inevitably make ideas difficult to understand.

3.1 CORRECT ALL SENTENCE FRAGMENTS.

A sentence fragment is a group of words, punctuated as a sentence, that is not a complete sentence. Sentence fragments are usually unclear and must be corrected for the writer to be totally effective. A fragment may be corrected in two ways.

1. *Turn the fragment into a subordinate clause.*

Fragment	Correction
Although he ran slowly.	Although he ran slowly, he finished second.
Because we were late.	Because we were late, the party was almost over.

2. *Turn the fragment into a complete sentence.*

Fragment	Correction
The two cowboys rode.	The two cowboys rode the range.
Gone from the office.	Many men are gone from the office today.

● **EXERCISE 2**   Correct each of the following sentence fragments.

1. Because I wanted to.
2. When I think.
3. Working for a small consulting firm.
4. Listening to the radio.
5. When in Silicon Valley.
6. Because of the computer.
7. After arriving at the office.
8. Since software came into my life.
9. When in the course of human events.
10. Whistling loudly and stomping his feet.

3.2   CORRECT ALL COMMA SPLICES.

A comma splice occurs when a comma instead of a period, semi-colon, or coordinating conjunction is placed between two main clauses. Unless the two clauses are very short and not possible to misread, such as "He came, he saw, he conquered," correct a comma splice by using one of several methods.

1. *Replace the comma with a period.*

Comma Splice	Correction
He walked away, I followed him.	He walked away. I followed him.

2. *Use a semicolon between the two clauses. This implies a closer connection than the use of a period.*

Comma Splice	Correction
It rained steadily, a rainbow followed.	It rained steadily; a rainbow followed.

3. *Insert a coordinating conjunction after the comma. Be sure to choose the conjunction that best fits the meaning of the sentence.*

Comma Splice	**Correction**
The plane was late, the meeting was cancelled.	The plane was late, and the meeting was cancelled.
Three people were ahead of me, I was chosen first.	Three people were ahead of me, but I was chosen first.

4. *Make one of the clauses subordinate.*

Comma Splice	**Correction**
Only one man wanted the chairmanship, three were considered ambitious.	Although three men were considered ambitious, only one man wanted the chairmanship.
Early computers used vacuum tubes, they often needed repairs.	Because early computers used vacuum tubes, they often needed repairs.

● **EXERCISE 3**   Correct each of the following comma splices.

1. The guest speaker left town early, the dinner was cancelled.
2. The manager resigned, two of his subordinates wanted the job.
3. Mosquitoes can cause harm, none compares to the diseases they transmit.
4. The meeting never occurred, I went to the wrong building.
5. Two out of the group resigned, the others remained.
6. I could invest in real estate, I could buy bonds.
7. There are three ways to solve the problem, two of them have already failed.

3.3   CORRECT ALL FUSED SENTENCES.

A fused sentence occurs when two main clauses are joined without any punctuation or without a coordinating conjunction. This error is sometimes called a run-on sentence. While not as common as a fragment or a comma splice, it is equally confusing. Correct fused sentences by the devices used to correct comma splices.

1. *Place a period between the clauses.*

Fused Sentence	**Correction**
The building collapsed the contractor was blamed.	The building collapsed. The contractor was blamed.

The plane left the runway the pilot relaxed.	The plane left the runway. The pilot relaxed.

2. *Place a semicolon between the two clauses.*

Fused Sentence	**Correction**
He made the sale he received his bonus.	He made the sale; he received his bonus.
The computer is down thus you should type your report.	The computer is down; thus you should type your report.

3. *Place a coordinating conjunction between the two clauses.*

Fused Sentence	**Correction**
The town was empty the bank was open.	The town was empty, but the bank was open.
The dorms closed the students went home.	The dorms closed, and the students went home.

4. *Make one of the clauses subordinate.*

Fused Sentence	**Correction**
The river was rising the town was flooded.	Because the river was rising, the town was flooded.
I was late again I kept my job.	Although I was late again, I kept my job.

Note: Correct all sentence errors by careful revision. Poorly written sentences are easily misunderstood by the reader.

●**EXERCISE 4**   Correct each of the following fused sentences.

1. The negotiations broke down each side blamed the other.
2. The country was threatened by its neighbors they were greedy.
3. There are two ways to do this you have chosen the wrong way.
4. No two people will see a painting in the same way each will see what he wants to see.
5. The company purchased a robot it can replace two humans.
6. Ghost towns have a certain charm they are very dirty.

●4

PRONOUNS

KEY POINTS

- Use the correct case for each pronoun.
- Be sure the pronoun and antecedent agree in number.
- Avoid ambiguous, broad pronoun references.
- Use a clear antecedent for each pronoun.
- Use *whom* not *who* in the objective case in formal writing.

A pronoun takes the place of a noun in a sentence, largely to avoid repetition and to allow for a more concise, clearer structure. Without pronouns, many sentences would bog down with excessive noun usage.

Without Pronouns	**With Pronouns**
When the foreman wanted more bricks, the foreman called the brickyard.	When the foreman wanted more bricks, *he* called the brickyard.
Jeff sold the computer and the computer's software as a package deal.	Jeff sold the computer and *its* software as a package deal.
Not only did the cars have bad transmissions, but the cars also had poor suspension systems.	Not only did the cars have bad transmissions, but *their* suspension systems were also poor.

4.1

4.1 USE THE CORRECT CASE FOR EACH PRONOUN.

"Case" refers to using the form of a pronoun that most clearly shows its function in the sentence. There are three cases: *subjective*, *possessive*, and *objective*. The function of each case is clear from its title. Pronouns in the subjective case are used as the subject of a sentence or as subjective complements—words that complete the meaning of a linking verb and modify or refer to the subject. Pronouns in the possessive case show ownership or a comparable relationship. Pronouns in the objective case are the object of a verb, verbal, or preposition, or the subject of an infinitive.

EXAMPLES

He was the last to finish the project. [subjective]
We hurried home after work. [subjective]
That was *she* typing furiously. [subjective used as subjective complement]
Don't borrow *my* new Word Perfect disk. [possessive]
The managers liked *his* presentation. [possessive]
Franklin and Carl found *her* ring under the table. [possessive]
John bought *himself* a new suit. [objective]
The foreman praised *me*. [objective]
The judge gave *us* five minutes to present our case. [objective]

The following chart gives the cases for personal pronouns:

	Subjective	Possessive	Objective
Singular			
First person	I	my, mine	me
Second person	you	your, yours	you
Third person	he, she, it	his, her, hers	him, her, it
Plural			
First person	we	our, ours	us
Second person	you	your, yours	you
Third person	they	their, theirs	them

●**EXERCISE 1**   Choose the correct case in each of the following sentences.

1. When the governor nominated the four judges, he said that *they/their* views agreed with *his/him*.

2. The board of directors reacted by demanding that *they/them* restate their views on the proposed merger.
3. Last week you and *I/me* were selected employees of the week.

4.2 USE THE POSSESSIVE CASE CORRECTLY.

While most writers recognize the need for the possessive case, many must learn it. They do not know its correct use and apply apostrophes unnecessarily. Apostrophes are not used with personal and relative (*who*) pronouns in the possessive case. But remember that indefinite pronouns, such as *anyone* or *everybody*, may indicate possession by -'s or by -s alone.

Incorrect	Correct
The job is your's.	The job is yours.
His' laziness cost him dearly.	His laziness cost him dearly.
The late shift exceeded it's production quota.	The late shift exceeded its production quota.
The project's failure is everybodys problem.	The project's failure is everybody's problem.

Note: Never write *it's* (it is) as a possessive. This is the most common error in pronoun usage. See Section 9.2, pp. 60–61, for further explanation.

●**EXERCISE 2**   Choose the correct form of the possessive case in each of the following sentences.

1. The job never lived up to *its/it's* expectations.
2. The increase in sales was *his/his'* doing.
3. The victory was *theirs/their's*.
4. The printers were *our's/ours*.

4.3 MAKE EACH PRONOUN REFER CLOSELY TO ITS ANTECEDENT.

Faulty pronoun reference is common and irritating to the reader. You can avoid reference errors by carefully editing each sentence for pronoun usage. A pronoun must refer to its antecedent, the word the pronoun is replacing, with absolute clarity. Reference errors are due largely to unclear thinking. If you understand the reference, you assume the reader also will. These are the two common reference errors:

1. Ambiguity

An ambiguous pronoun reference forces the reader to choose between two possible antecedents.

EXAMPLES

Torn between Tom's invitation and Harry's request, Dick chose *his* offer. (*His* should be changed to either *Tom's* or *Harry's*.)

Paul called to Luis as *he* turned the corner. (*He* should be changed to either *Paul* or *Luis*.)

2. Remote or broad references

A pronoun should be placed as close as possible to the antecedent. The more distance you put between the two, the more room for misreading.

EXAMPLES

Some companies see retirement in 1986 as a way to trim the labor force; the workers see it as a way to leave a job before they are too old. Whatever the cost, both labor and management must learn to use *this* creatively. (*This* is a very remote reference to *retirement*. Replace *this* with *retirement*.)

Susan wanted her supervisor to rate her work as highly as he would have done if she had done *it* carefully. (*It* should be replaced by *her work*.)

Note: Even though readers might recognize the antecedent after a few moments, do not make them do the writer's work. There are few demands more certain to irritate readers.

●**EXERCISE 3**   Identify faulty pronoun references in the following sentences. Correct each one.

1. Jane called Mary every day when she was free.

2. The president met with his staff at the end of the fiscal year. After talking to the managers, foremen, and other members of the company, they decided to increase production.
3. Ramon refused to attend the noisy sales conference, which angered some of the marketing team.
4. The agreement between management and the union took effect when they invited them to lunch.
5. Whenever Lance met John, he greeted him.
6. Because the setting was incorrect and the metal sheets too thick, the stamping machine broke down after trying to stamp it.

4.4 USE THE IMPLIED *YOU* CORRECTLY.

When using the imperative mood, or implied *you*, remember that this mood gives commands and is generally used in writing instructions.

EXAMPLES

Read this memo.
Bring me the blueprints.
Run from the fire!

4.5 WHEN USING *IT* AS A PERSONAL PRONOUN, WHETHER TO REFER TO A SPECIFIC NOUN OR AS AN EXPLETIVE (*IT* IS COLD THIS MORNING), MAKE SURE THE REFERENCE IS CLEAR.

Vague	Improved
Although it was snowing on the street, *it* was not impassible.	Although it was snowing on the street, *the walk* was not impassible.
	or
	The snow did not make the street impassible.
When it was time to leave, *it* had struck twelve.	When it was time to leave, *the clock* had struck twelve.

4.6 AVOID USING *THEY* OR *THIS* WITH A BROAD AND UNCLEAR ANTECEDENT.

Pronoun references should be clear and specific, not implied or broad. This is particularly true with the common pronouns *they* and *this*.

Vague	Improved
When the factory opened, *they* went to work.	When the factory opened, *the workers* arrived for the first shift.
The president asked for a few good men, but *they* were not impressed.	The president asked for a few good men, but *his audience* was not impressed.
Carl Swenson said the factory would remain open. *This* made the workers happy.	Carl Swenson said the factory would remain open. *This confirmation* made the workers happy.

4.7 IN FORMAL WRITING, OR TO BE ABSOLUTELY SURE OF PRECISENESS, ALWAYS USE *WHOM* INSTEAD OF *WHO* FOR OBJECTS OF A SENTENCE.

Informally, *who* is often substituted for *whom;* however, *whom* should be used in the objective case in formal writing. The confusion of *who* and *whom* can be easily avoided if you make certain of the pronoun's function in the sentence.

EXAMPLES

Whom as Object	Who as Subject
For *whom* are the men looking?	The foreman, *who* is lazy, will be replaced.
Whom did you say you wanted?	I wanted the painter *who* was the least expensive.

Note: The troublesome choice between *who* and *whom* can be avoided by careful rewriting without the *who/whom* construction.

4.7

●**EXERCISE 4**   Identify the correct pronoun or clarify the pronoun in each of the following sentences.

1. The blizzard made it difficult for it to be visible.
2. The manager said he would use a five-day rotation, but they were not convinced.
3. *Who/Whom* is happy?
4. He is the man *who/whom* we are seeking.
5. He is the man *who/whom* is our leader.
6. There is no one *who/whom* we would hire without a thorough interview.

●5

ADJECTIVES AND ADVERBS

KEY POINTS

- Do not overload sentences with adjectives and adverbs.
- Be sure adjectives modify nouns or pronouns and adverbs modify verbs, adjectives or adverbs.
- Hyphenate with adjectives when appropriate.
- When comparing, use the comparative with two subjects and the superlative with more than two subjects.

The overuse of adjectives and adverbs can clutter and obscure any piece of writing. However, if used carefully, both adjectives and adverbs make ideas clearer and more specific. To write effectively, you must learn to distinguish between adjectives and adverbs and must become aware of the basic problems in their usage.

5.1 USE AN ADJECTIVE TO MODIFY NOUNS OR PRONOUNS.

a. Normally, place adjectives directly before the word they modify.

EXAMPLES

The *ambitious* salesperson took on *extra* territory.
The *early* impact of the *emerging* technology was in the field of medicine.

5.2

The *smaller* subsidiaries in the United States are part of the *umbrella* company.

b. Use an adjective as the complement of a subject or an object.

A complement is a word, often an adjective, that modifies or refers to the subject (subject complement) or object (object complement).

EXAMPLES

The plans make the building *modern*. [object complement]
The players found the coach's game plan *exciting*. [object complement]
The plans are *strange*. [subject complement]
The novel is *complex*. [subject complement]

Note: The subject complement is also referred to as a predicate complement.

5.2 USE AN ADVERB TO MODIFY VERBS, ADJECTIVES OR OTHER ADVERBS.

Adverbs should be used carefully. However, without adverbs, the writer would find it much more difficult to give the exact meaning of the sentence.

Original	Modified
The market research firm compiled the figures.	The market research firm compiled the figures *inaccurately*.
Their share of the sales dropped during the last quarter of 1989.	Their share of the sales dropped *sharply* during the last quarter of 1989.

Note: One major problem with adverbs is the occasional confusion between formal and informal usage.

Informal	Formal
The new computer *sure runs good*.	The new computer *surely runs well*.
The stone depicts the Aztec cosmos *just perfect*.	The stone depicts the Aztec cosmos *perfectly*.

Note: One of the most common mistakes is the use of *real* as an adverb. Observe what it modifies.

Faulty	**Improved**
He was *real* happy to be asked to speak at the conference.	He was *very* happy to be asked to speak at the conference.
Wayne had a *real* good job.	Wayne had an *extremely* good job.

Note: *Real* can be used only as an adjective.

EXAMPLE

He was a *real* leader.

5.3   WITH VERBS RELATING TO THE SENSES—*FEEL, LOOK, SMELL, SOUND, TASTE*—CHOOSE ADJECTIVES OR ADVERBS APPROPRIATE TO THE SENTENCE PART BEING MODIFIED.

EXAMPLES

The sewage disposal area smells *bad.* [adjective modifying area]
The sewage disposal area smells *badly.* [adverb modifying *smells*]
John looked *happy* when he was promoted. [adjective modifying *John*]
John looked *happily* at his letter of promotion. [adverb modifying *looked*]
The bell sounds *loud.* [adjective modifying *bell*]
The buzzer sounds *clearly.* [adverb modifying *sounds*]

5.4   USE HYPHENS WITH ADJECTIVES WHEN APPROPRIATE.

a. If two or more words form a single adjective before a noun, hyphenate the words.

EXAMPLES

He bought a *Hayes-compatible* modem.
It was a *record-setting, end-of-the-year* sale for the *audio-video* company.

5.5

b. Hyphenate numbers spelled out from one to nine and figures from 10 on when using them as adjectives.

EXAMPLE

The *nine-inch diameter* tube was too small.
A *55-year-old* movie may have strange dialogue.

Weak	Improved
The *long range* plans were vague.	The *long-range* plans were vague.
He was offered a *no cut* contract.	He was offered a *no-cut* contract.

Note: Normally hyphenate compound words or numbers combined with words before nouns, not after them.

Before a Noun	After a Noun
The *day-old* formula was flat.	The flat formula was a *day old*.
Their *10-year* contract expired.	Their contract expired after *10 years*.

●**EXERCISE 1**   Choose the correct adjective or adverb in each of the following sentences.

1. He *sure/surely* programs fast.
2. Cassettes were a *real/really* bad choice.
3. My superior was *real/really* angry because I was late for the meeting.
4. The situation was *very/sure* uncertain, and we responded to the danger *slow/slowly*.
5. He gave a *masterful/masterfully* response.
6. We spent three weeks *lazy/lazily* at a cabin in the mountains.
7. "I believe you have failed," he said *angry/angrily*.

5.5   USE THE COMPARATIVE AND SUPERLATIVE DEGREES CORRECTLY WHEN COMPARING BOTH ADJECTIVES AND ADVERBS.

In general, use the comparative degree when comparing two persons or things, and the superlative degree when comparing three or more persons or things. Shorter adjectives form the comparative by adding *-er* and the superlative by adding *-est*.

EXAMPLES

The X982 is *larger* than the X863.
The X982 is *the largest* computer in the series.

Adverbs and most adjectives of three or more syllables form the comparative by using *more/less* and the superlative by using *most/least*.

EXAMPLES

His new job has *more* pressure than his old one did.
His new job has *the most* pressure of any of his jobs.
Sarah has *less* seniority than Janet.
Sarah has *the least* seniority of any of the chemists in the soils lab.

Note: Do not use a double comparative or superlative.

Weak	Improved
The Monday night shift was *more lazier* than the Tuesday day shift.	The Monday night shift was *lazier* than the Tuesday day shift.
She was the *most strongest* of the team members.	She was the *strongest* of the team members.

●**EXERCISE 3**   Correct the errors in the use of the comparative and superlative in the following sentences.

1. Tracy was the older of the programmers.
2. Rose hips are the better source of Vitamin C.
3. Oak has the more BTUS than pine.
4. The giant spider crab of Japan is the larger of all other crustaceans.
5. Joan's cover had the most appealing qualities of the two entries for adoption by the graphics department.

●6

SHIFTS

KEY POINTS

- Check each sentence twice: once for verb shifts and once for pronoun shifts.
- Write all directions in the imperative mood.
- When using collective nouns, check nouns and verbs for shifts in tense or number.

Needless shifts in tense, person, voice, number, or mood within a sentence will make writing unclear and irritating to the reader. These shifts are a problem because they are not always easy to detect and may occur in a variety of situations.

6.1 DO NOT SHIFT TENSE UNNECESSARILY.

Almost without exception, all verbs in a sentence or in a paragraph should be in the same tense. Shifting tense for no reason makes it difficult for your reader to determine when the matter under consideration is taking place, has taken place, or will take place.

EXAMPLES

Shifts from Past to Present	Revisions
After arriving at work, the reporter questioned the placement of his	After arriving at work, the reporter questioned the placement of his

story, but his editor replies that he will make those decisions.

story, but his editor replied that he would make those decisions.

I called my broker to place an order, but he tells me the stock is too risky.

I called my broker to place an order, but he told me the stock was too risky.

● **EXERCISE 1**   Identify and correct the unnecessary shifts in tense in the following sentences.

1. In the nineteenth century, few cowboys had a formal education, and those who did often do not mention the fact.
2. In the beginning, the job was excellent but, in a few weeks, the conditions worsen.
3. The family business, originating in 1870, makes progress with a new plant in 1935.
4. When he arrives for work, he learned what his assignment for the day will be.
5. Some day people are traveling into space and return to earth on regular space-shuttle flights.
6. The first computer, which is built in the 1940s, used vacuum tubes to process information.
7. Geothermal systems share one thing in common—a natural heat source, the natural radioactivity that existed in rocks.
8. The Battle of Gettysburg is fought in 1863; General Meade was the Union commander.

6.2 DO NOT SHIFT PERSON OR POINT OF VIEW UNNECESSARILY.

Most writers use the third person because it is less personal. However, whether you use first person (*I*), second person (*you*), or third person (*he* or *she*), you should use the same point of view throughout your work.

Note: Point of view refers to whoever is speaking in a piece of writing.

6.4

EXAMPLES

Shift from Third to Second Person

They were intimidated by the MS-DOS that appeared on the screen when you turned on the computer.

Third Person Throughout

They were intimidated by the MS-DOS that appeared on the screen when they turned on their computers.

Shift from First to Third Person

When I look at the choices available, they have little hope of finding a solution.

First Person Throughout

When I look at the choices available, I have little hope of finding a solution.

6.3 DO NOT SHIFT FROM THE ACTIVE TO THE PASSIVE VOICE.

It is legitimate to shift from active to passive voice from one sentence to the next, but not within a sentence. See Section 26, pp. 164–66, for more information on voice.

Awkward shift

The manager approved the plans, but they were not liked by him.

When the indexes were being moved by John, a chair fell on him.

No Shift

The manager approved the plans, but he did not like them.

When John was moving the indexes, a chair fell on him.

Note: Whenever possible, use the active voice, as it usually makes material more interesting.

6.4 DO NOT SHIFT BETWEEN SINGULAR AND PLURAL SUBJECTS AND VERBS OR PRONOUNS AND ANTECEDENTS; MAINTAIN AGREEMENT.

Such errors are among the most common and most confusing mistakes a writer can make.

No Agreement	Agreement
Each series of the negotiations had their problems.	Each series of the negotiations had its problems.
Everybody needs to keep track of their appointments.	Everybody needs to keep track of his or her appointments.
The players have their plans for the war games and is ready for the tournament.	The players have their plans for the war games and are ready for the tournament.
BASIC and COBOL is two computer languages he knows and use often.	BASIC and COBOL are two computer languages he knows and uses often.

6.5 DO NOT SHIFT MOOD WITHOUT A VERY GOOD REASON.

A writer who shifts from the indicative to the imperative or to the subjunctive mood is asking to be misunderstood.

Shift	No Shift
Pick up the boxes and then you should carry them to the other room.	Pick up the boxes and carry them to the other room.
I will change three things if I were in control.	I would change three things if I were in control.

Note: When giving directions, rely on the imperative mood for conciseness and clarity.

EXAMPLES

Turn left at the next corner and drive for three miles.
Rotate the dial for three turns.

●**EXERCISE 2**   Correct any shift errors in the following sentences.

1. Although Einstein first thought of the theory used in lasers as early as 1917, the first laser was built in 1960 by Theodore H. Maiman.

2. When one considers how to reopen a business, you must be very careful.
3. In early computers the vacuum tube gave off vast amounts of heat and was burned out often.
4. No one made much of their abilities.
5. Sign the checks and then you should talk to me.

PUNCTUATION

●7

COMMAS

KEY POINTS

- Use commas whenever necessary to prevent misreading of a sentence.
- Do not overuse commas—make every comma count.
- Use commas when a pause is necessary.

Of all the punctuation marks, commas are the most misused. Even one misused comma can cause misreading of a sentence. If a writer thinks of the comma as the sign of a brief pause, its use will be clearer. Reading a sentence aloud will often indicate where commas should be placed. Because of increased technical accuracy in documents, today it has become more important than ever to use the comma correctly.

7.1 ALWAYS PUT A COMMA BEFORE A COORDINATING CONJUNCTION THAT JOINS TWO MAIN CLAUSES.

The coordinating conjunctions and their placement between two main clauses are as follows: *and, but, or, so, for, nor, yet.*

MAIN CLAUSE comma *yet*   MAIN CLAUSE

Weak	**Improved**
The manager called a meeting for three and anyone not attending was fired.	The manager called a meeting for three, and anyone not attending was fired.

The flight paths merged but the air traffic controllers were unaware of the situation.

The flight paths merged, but the air traffic controllers were unaware of the situation.

The business declared bankruptcy yet the partners remained friends.

The business declared bankruptcy, yet the partners remained friends.

7.2 DO NOT USE A COMMA BEFORE A COORDINATING CONJUNCTION JOINING TWO WORDS OR PHRASES.

Weak	Improved
The database was useful, but not the total answer.	The database was useful but not the total answer.
The outlook for the plant expansion was optimistic, but uncertain.	The outlook for the plant expansion was optimistic but uncertain.

●**EXERCISE 1**   Add necessary commas and delete unnecessary commas in the following sentences.

1. We were here early but no one was present.
2. Two young men moved the copier from the workroom, and set it in the main office.
3. She will not be welcome nor will she be allowed to participate in the next teleconference.
4. The best that can be said of Leo is that he is prompt, but lazy.
5. It is neither breakable nor bendable.
6. The programs had come a long way yet they were finished.
7. Hank made the sale so he asked for the commission.

7.3 PUT COMMAS AFTER INTRODUCTORY ADVERB CLAUSES, PHRASES, INTERJECTIONS, AND THE WORDS YES AND NO WHEN USED AT THE BEGINNING OF A SENTENCE.

Adverb Clauses   An adverb clause is a dependent clause that functions as an adverb in a sentence.

EXAMPLES

> *When it was time,* we left for the conference.
> *Because John was late,* he was fined.
> *Although Joseph thought Acme Products offered the best terms,* he did
>   not give them the contract.

Phrases A phrase is a group of words that are related but do not include both a subject and a verb.

EXAMPLES

> *In most situations,* nitroglycerin will react predictably.
> *Displaying consistent taste throughout the years,* I have disliked the
>   line-printer font.
> *Overriding my fear of calculus,* I chose engineering as an major.

Interjections An interjection is a word expressing a simple exclamation.

EXAMPLES

> *Oh,* we should buy some new electroplating equipment.
> *Whew,* it must be 400 degrees in there.

Introductory Words The words *yes* and *no* at the beginning of a sentence are treated the same as interjections.

EXAMPLES

> *Yes,* I am prepared for the meeting.
> *No,* we will not adopt the new operating system.

Note: If uncertain whether to use a comma after an introductory element, test the sentence to see whether or not it can be understood clearly without the comma.

7.4 USE THE COMMA CORRECTLY WITH A SERIES OF THREE OR MORE ITEMS.

Usually use a comma after each item in a series but the last one. The commas prevent misreading and emphasize that each item in the series is separate and does not modify the next word in the series. When the last item in a series is separated from the other items by a coordinating conjunction, the preferred usage is to place a comma before the coordinating conjunction.

EXAMPLES

The building was tall, dark, and ugly. (The comma before *and* makes misreading impossible.)

The building was tall, dark, ugly. (Without the coordinating conjunction, a comma after *dark* prevents misreading.)

The building was tall and dark and ugly. (Commas are unnecessary since the use of *and* prevents misreading.)

It was a tall, dark, ugly building. (If the items in the series are being used to modify one word in the sentence, they should be separated with commas.)

●**EXERCISE 2**   Insert necessary commas in the following sentences.

1. Yes we have no printers.
2. The office decor was loud bright and ugly.
3. The electric kiln was small and old and dangerous.
4. Because the disk was full Louise formatted another.
5. The job was hard dangerous and well-paying.
6. No I will not join the local professional chapter.
7. Although my time is valuable I will reprogram the robots for you.
8. The new manager was forceful demanding and hard working.
9. Ouch I pinched my finger!
10. Whenever I feel depressed I play computer games.

7.5 SET OFF NONRESTRICTIVE ADJECTIVAL CLAUSES AND PHRASES WITH COMMAS.

A nonrestrictive clause or phrase does not restrict the meaning of the noun or pronoun being modified; it is not necessary for identifying the modified word.

EXAMPLES

John Huston directed *The Maltese Falcon*, which has one of Humphrey Bogart's greatest performances. (Information about Bogart is not essential for identifying the film.)

John Huston, Walter Huston's son, directed *Prizzi's Honor.* (John Huston's father is not essential for identifying the director.)

7.6 DO NOT PLACE COMMAS AROUND A RESTRICTIVE PHRASE OR CLAUSE THAT LIMITS THE MEANING OF A SENTENCE.

With commas, the sentence will not be clear to the reader.

Weak	Improved
The genius, who directed *The Maltese Falcon,* is appearing on television.	The genius who directed *The Maltese Falcon* is appearing on television.
The woman, who led the relief drive, was honored at a luncheon.	The woman who led the relief drive was honored at a luncheon.

Note: In the above sentences, the genius is not specifically identified when "who directed *The Maltese Falcon*" is enclosed in commas. Without commas, the genius is identified. The woman who is being honored at a luncheon is unidentified if her role in the relief drive is set off by commas.

●**EXERCISE 3**   Insert necessary commas in the following sentences.

1. The transistor Bell Telephone's finest contribution to technology was invented 30 years ago.
2. President Roosevelt the leader of his country declared a bank holiday.
3. The man referred to as Slick was the founder of our company.
4. One of America's greatest books *Huckleberry Finn* is still banned in some high schools.
5. The union's contract which expired in August was renegotiated over a lengthy period of time.
6. The fast-food restaurants that moved to town at the same time were highly competitive.

7.7 SET OFF PARENTHETICAL EXPRESSIONS WITH COMMAS.

Parenthetical expressions are words, phrases, or clauses inserted in a sentence that are not absolutely necessary or a part of the main thought.

EXAMPLES

It was clear, indeed, that the office manager was efficient. (*Indeed* is unnecessary but does add intensity.)

Geraldine smiled, although few saw her, when the speaker was introduced. (*Although few saw her* is a parenthetical clause.)

I told you, John, to be careful of the drill. (*John* is a direct address used parenthetically.)

I thought, of necessity, that we should act. (*Of necessity* is a parenthetical prepositional phrase.)

7.8 USE COMMAS WHENEVER NECESSARY TO PREVENT MISREADING OF A SENTENCE.

Unclear	Clear
Junior executives who like to impress their bosses are often the last to leave work.	Junior executives, who like to impress their bosses, are often the last to leave work.
On a farm yards are sometimes pastures.	On a farm, yards are sometimes pastures.

7.9 USE COMMAS CORRECTLY WITH QUOTATION MARKS.

a. Place commas within quotation marks.

Weak	Improved
"I regret this decision", the president said.	"I regret this decision," the president said.
"We have come a long way", the space technician said.	"We have come a long way," the space technician said.

7.10

b. Use commas to separate unquoted material from direct quotes.

Weak	Improved
"John," he said "I want to resign as of today."	"John," he said, "I want to resign as of today."
"Marie" her superintendent said "you must finish your report by noon today."	"Marie," her superintendent said, "you must finish your report by noon today."

● **EXERCISE 4**   Whenever necessary, insert commas in the following sentences or fix incorrectly placed commas.

1. "I could continue the program", the director said.
2. "Helen" my manager would always say "either you're a top sales representative or I'm misreading your reports."
3. There are those who believe it would have been better for Rhett Butler to have said", Frankly my dear, I don't give a darn!"
4. "I have only one reply to your offer of merger" the president stated although he thought otherwise.
5. "I will stand by my original projection" the engineer replied.
6. I believe however that the delay is unnecessary.
7. Frank fidgeted in his chair during the meeting although he appeared calm but only one person noticed.
8. Two of the analysts for good reason decided to look elsewhere for the cause.
9. When I talk to you Estelle I expect you to listen.
10. After touring the plant robots were all he ever talked about.

7.10 USE COMMAS CORRECTLY WITH DATES, ADDRESSES, GEOGRAPHICAL LOCATIONS, AND TITLES.

a. Use commas after the day of the month, the date, and, if within a sentence, after the year.

EXAMPLES

We received the order on September 11, 1989, and filled it immediately.
We received the order on Tuesday, September 12, 1988.

b. If the specific date is omitted, do not use a comma between the month and the year or after the year.

EXAMPLE

The factory was opened in May 1945 and has been in full production ever since.

c. Use no commas if writing a date in this order: day-month-year.

EXAMPLE

25 July 1989

d. Use commas with addresses after both the city and state if both occur within the sentence.

EXAMPLES

Platteville, Wisconsin, is a city in the southwestern part of the state.
I will move to Scottville, Michigan.

Note: Do not use a comma before a zip code.

Weak	Improved
Stacy's new address is 300 West Fifth Street, Dubuque, Iowa 52001.	Stacy's new address is 300 West Fifth Street, Dubuque, Iowa 52001.

e. Use a comma to separate names of cities from names of states or countries.

EXAMPLES

Leo wanted to negotiate foreign contracts, but he spent only two weeks in Paris, France, and three months in Brussels, Belgium, where he had originated agreements, before he returned to Attica, New York.
When a Chinese friend asked me what American cities to visit, I told him Ames, Iowa, and Lincoln, Nebraska.

f. If a title follows a proper name, separate them with a comma.

EXAMPLES

Charles Mayo, MD
Jacob Williams, President

●**EXERCISE 5**   Insert commas where they are necessary. Delete any unnecessary commas.

1. Pumpkin Center South Dakota is a town that time forgot.
2. The company's new address is 1515 Duck's Breath Lane Metropolis Minnesota, 77513.
3. He was born in December, 1933 and shortly afterward his family was transferred from Pierpont South Dakota to Spotted Horse Wyoming.
4. His term of office will expire on 1, January, 1988.
5. F. O. Matthiessen PHD and Milton Zulnick EDD found too little to agree about when discussing Melville's poetry at a conference in May, 1937.
6. During one summer, I lived in Pippa Passes, Kentucky and, for a few days, in Dayton, Ohio.
7. A date most Americans will not soon forget is November 22 1963.
8. The corporate office's address is 1 Grandview Drive Pocatello Idaho.
9. Elmer Lemon DDS has been in practice for 30 years.
10. Two of the smallest state capitals are Pierre South Dakota and Carson City Nevada.

The basic rules for comma use are summarized in the following chart:

Do Not Use	Use
Before coordinating conjunctions that link only two words or phrases.	Before coordinating conjunctions used to link two main clauses.
After short introductory phrases with no chance of misreading.	After introductory adverb clauses.
	After *yes, no,* interjections, and transitional words, such as *certainly* or *on the other hand.*
With words in a series separated by coordinating conjunctions.	With two or more words in a series.
With restrictive phrases and clauses.	With nonrestrictive phrases and clauses.
With very short parenthetical expressions.	With parenthetical expressions that clarify the word being modified.
When there is no possibility of misreading, even with dates.	To prevent misreading.
	Within direct quotes and to separate quoted from nonquoted material.
With dates that reverse the month and day.	To separate items in dates, addresses, geographic locations, and cities.

●8

SEMICOLONS

KEY POINTS

- Use a semicolon instead of a period when two main clauses are closely linked.
- Do not join elements with a semicolon that are of unequal importance.
- Do not use a semicolon when a colon is required.

While not used as heavily as the comma, the semicolon is important. It is sometimes referred to as the "middle ground" between a period and a comma or between a brief pause and a complete pause. Used with discrimination, the semicolon will make your writing clearer and more interesting.

8.1 USE A SEMICOLON TO SEPARATE TWO MAIN CLAUSES NOT SEPARATED BY A COORDINATING CONJUNCTION (SEE SECTION 3.2).

EXAMPLES

The foreman changed the work schedule; the workers were angry.
Trains run on time in Europe; in America they sometimes run.

Note: As mentioned in Section 3.2, using only a comma between two main clauses will produce a comma splice. Always remember that the semicolon distinguishes conjunctive adverbs from coordinating conjunctions.

8.2

EXAMPLES

The plant manager accepted the unconditional terms, but he was dissatisfied. [coordinating conjunction]
The plant manager accepted the unconditional terms; nevertheless, he was dissatisfied. [conjunctive adverb]

8.2 USE A SEMICOLON BEFORE A CONJUNCTIVE ADVERB OR A TRANSITIONAL WORD OR PHRASE TO SEPARATE TWO MAIN CLAUSES.

Frequently used conjunctive adverbs and transitional phrases are listed below:

Conjunctive Adverbs

also	however	nonetheless
anyway	incidentally	otherwise
besides	indeed	still
consequently	likewise	then
furthermore	meanwhile	therefore
hence	nevertheless	thus

Transitional Phrases

after all	even so	in the second place
as a result	for example	on the contrary
at any rate	in addition	on the other hand
at the same time	in fact	
by the way	in other words	

EXAMPLES

Barton thought he was a great engineer; however, he had never mastered chemistry. [MAIN CLAUSE; conjunctive adverb, MAIN CLAUSE]

Barton thought he was a great engineer; indeed, he might have been if he had ever mastered chemistry. [MAIN CLAUSE; conjunctive adverb, MAIN CLAUSE]

Barton thought he was a great engineer; of course, he might have been if he had only mastered chemistry. [MAIN CLAUSE; transitional phrase, MAIN CLAUSE]

Note: A period may be used in place of the semicolon, creating two sentences.

EXAMPLE

> Barton thought he was a great engineer. Indeed, he might have been if he had ever mastered chemistry. [conjunctive adverb]

Note: It is possible to place a semicolon before a coordinating conjunction, either to indicate a contrast between the two clauses or to separate two long clauses.

EXAMPLES

> Throughout history, many people have disliked insecticides; but they have used them on their crops because of economic necessity. Throughout the past century, many technological advances have reached into every nook and cranny of modern life; and none has been more prominent and more far-reaching than the one we take most for granted—the automobile.

● **EXERCISE 1**   Insert necessary semicolons and correct unnecessary semicolons in the following sentences.

1. He made only two designs, then he quit.
2. Maria accepted the company's offer however; she did add a few stipulations.
3. Stevens always brought his projects in under budget furthermore he won more contracts.
4. There are those who don't believe a report should be kept; but they have never tried to locate last year's budget.
5. Leo would not accept our figures, he believed we could buy the machinery for a lower price.
6. The board of directors was elated about winning the proxy battle, but the major stockholders were dejected about losing the battle.
7. The IRA rejected several of my tax exemptions, furthermore they billed me for three years' interest.
8. The Battle of the Bulge was Germany's last attempt to snatch victory from defeat, certainly, most historians would agree with this.
9. Will Rogers never met a man he didn't like, he never met my boss.

8.3   USE SEMICOLONS TO SEPARATE ITEMS IN A SERIES WHEN THE ITEMS ALREADY CONTAIN COMMAS.

Weak	Improved
While reorganizing the office, Fred found old reports, a stack of memos, faded, bent photographs, and a box of cancelled checks.	While reorganizing the office, Fred found old reports; a stack of memos; faded, bent photographs; and a box of cancelled checks.
Breaking into King Tut's tomb, the adventurers uncovered broken, discolored, crude pottery, decorative, ornate jewelry, and, of course, an ebony sarcophagus.	Breaking into King Tut's tomb, the adventurers uncovered broken, discolored, crude pottery; decorative, ornate jewelry; and, of course, an ebony sarcophagus.

Note: If only commas are used in the above sentence, a misreading is quite likely.

8.4   DO NOT SEPARATE GRAMMATICALLY UNEQUAL PARTS OF A SENTENCE WITH A SEMICOLON.

a. Do not separate a phrase and a clause with a semicolon.

Weak	Improved
Where was Frank; the always nasty custodian?	Where was Frank, the always nasty custodian?
Of all things; he wanted to study French architecture.	Of all things, he wanted to study French architecture.

b. Do not separate a subordinate clause and a main clause with a semicolon.

Weak	Improved
Because Frank was late; the wastebaskets were full.	Because Frank was late, the wastebaskets were full.
Although we wanted to go to the conference; we went to the electronics exhibit instead.	Although we wanted to go to the conference, we went to the electronics exhibit instead.

8.5

8.5 DO NOT USE A SEMICOLON INSTEAD OF A COLON.

a. Do not use a semicolon instead of a colon at the end of a salutation in a business letter.

Weak	Improved
Dear Mr. Green;	Dear Mr. Green:

b. Do not use a semicolon between two main clauses when the second clause restates the first.

Weak	Improved
The study was a failure; it ended in disaster.	The study was a failure: it ended in disaster.
Richard had an accident; he hit a truck with a company car.	Richard had an accident: he hit a truck with a company car.

Note: For a discussion of the colon, see Section 11.4, p. 71.

●**EXERCISE 2** Delete unnecessary semicolons in the following sentences and replace them with correct punctuation.

1. After he retired, Buster Pilsner presented his mementos to the Hall of Fame: fifty dirty, torn, autographed baseballs, a nearly worn-out, faded, decrepit uniform, several tarnished, indecipherable plaques, and the bat he used to get his three-thousandth hit.
2. My boss has a predictable personality; always grumpy.
3. When I was nineteen; I decided I wanted to travel.
4. Dear Mr. Fuller;
5. The comptroller admitted that he had been siphoning money into a Swiss bank account; he admitted to being an embezzler.
6. After seeing *Close Encounters of a Third Kind;* one can have greater hope for the future of mankind.
7. Camillo Rodriguez was the most valuable employee the company had ever hired.

8.5

8. One or two people were present at the conclusion; and they were asleep.

9. My Uncle Luis left me a great deal in his will: 50 worn, cut-up, mildewed copies of the *National Geographic*, 10 broken, irreparable tin watches, and 500 canceled postage stamps.

10. Although Pete wanted to graduate; he was 10 credits short.

●9

APOSTROPHES

KEY POINTS

- Be sure to distinguish between *'s* and *s'* when forming possessives.
- Use apostrophes with contractions.
- Do not confuse *it's* and *its* or *whose* and *who's*.
- In special cases, use apostrophes to prevent misreading.
- Do not use apostrophes with personal pronouns.

Apostrophes are used to form possessives, contractions, and, in a few cases, plurals. They are commonly misused when a writer is dealing with possession or forming plurals. These errors would not occur if the writer would think about how apostrophes are being used. Avoiding such errors is extremely important since they are very noticeable.

9.1 USE THE APOSTROPHE WITH NOUNS AND INDEFINITE PRONOUNS TO INDICATE POSSESSION.

a. With singular nouns and indefinite pronouns, use an apostrophe followed by s.

EXAMPLES

Beverly's computer	anyone's car
engineer's plan	company's policies

9.1

b. With plural nouns ending in *s*, use only an apostrophe.

EXAMPLES

bosses' ideas	companies' policies
workers' hours	employees' wages

c. With plural nouns not ending in *s*, use an apostrophe followed by *s*.

EXAMPLES

men's equipment	people's choice
children's hour	mice's cages

d. In compound formations, use an apostrophe and *s* only with the last word.

EXAMPLES

chairwoman of the board's report
owner of the company's ideas
administrative assistant to the president's report

e. Where ownership by two or more people is not shared, use an apostrophe with each noun to indicate possession.

EXAMPLES

the foreman's and the lawyer's views
Bill's and Joe's lockers
men's and women's progress reports

f. When ownership by two or more people is shared, use the possessive with only the last noun.

EXAMPLES

Bill and Marcia's project
her aunt and uncle's office
the engineer and drafter's business

g. Do not confuse 's, s', or s's at the end of people's names.

If more than one person's name ends in *s*, do not confuse the *s* with the possessive. Use *es'* to show the possessive. If more than one person's name does not end in *s*, use *s'* to show the possessive. If one person's name ends in *s*, use *'s*.

Weak	Improved
the Jones's	the Joneses'
the Goldberg's	the Goldbergs'
Thomas' briefcase	Thomas's briefcase

● **EXERCISE 1** Whenever necessary, add apostrophes to indicate possession, or delete *s*.

1. The worlds greatest hope is nuclear fusion.
2. What were they planning for the companys next takeover?
3. Janes and Jeans work stations were next to one another for six years.
4. The assistant managers credentials were outstanding.
5. Lens and Jims project was postponed for 10 months of the year.
6. Many of Mark Twains works were published after his death.
7. The 20 employees reaction to the salary cut was predictable.
8. The chief engineers opinion was highly valued.
9. The workers and the managers views were different.
10. I thought this was exclusively the companys club.

9.2 USE AN APOSTROPHE TO INDICATE LETTERS OMITTED IN CONTRACTIONS.

EXAMPLES

He isn't here.
They're coming.
There's the building.
It is twelve o'clock (instead of *of the clock*).

Note: The confusion between contractions and the possessive causes two of the most common errors in English usage: the confusion of *its* and *it's* and, to a lesser degree, the confusion of *whose* and *who's*.

9.5

EXAMPLES

> This is its proper place. [possessive]
> It's being moved to its proper place. [contraction and possessive]
> We know whose calculator has been lost. [possessive]
> We know who's next on the program. [contraction]

9.3   DO NOT USE THE APOSTROPHE WITH A PERSONAL PRONOUN (I, YOU, HE, SHE, IT, WE, THEY) TO SHOW POSSESSION.

EXAMPLES

Weak	**Improved**
The next move is yours'.	The next move is yours.
His' house is next door.	His house is next door.

9.4   USE AN APOSTROPHE TO INDICATE OMISSIONS IN NUMBERS.

EXAMPLES

Vintage '76   Class of '56

9.5   IN A FEW CASES, USE AN APOSTROPHE TO FORM PLURALS.

a. Use an apostrophe to form the plural of lowercase letters to avoid confusion.

EXAMPLES

his f's	1900's
her d's and c's	10's

Note: In recent years the use of the apostrophe has become optional. Using it is up to the writer.

ALSO ACCEPTABLE

his fs	1900s
her ds and cs	10s

9.5

b. Use an apostrophe to form the plural of abbreviations followed by periods.

EXAMPLES

> Be sure to add the *Ibid.*'s.
> The *J.P.*'s convention reassembled

c. If misreading is a possibility, use an apostrophe to form plurals of capital letters and words labelled as words.

EXAMPLES

> He wrote too many *E*'s in this letter.
> Add some *I*'s to the note.
> After three or four *no no*'s, the child stopped fussing.

●**EXERCISE 2**   Insert necessary apostrophes in the following sentences.

1. You cant be right.
2. Formula 702s properties are dangerous and powerful.
3. He was told to mind his ps and qs.
4. Leroy added glamour to the class of 28.
5. I know whose car this is, but hes working in another plant now.
6. Watch for a fur watchband; its the newest fad.
7. He told me he was receiving too many Cs in his classes.
8. Benjamin Franklin was born in the 1700s.
9. They want to advance the deadline, but lets not allow them to do so.
10. This wasnt the time to use our new tool.

●10

QUOTATION MARKS

KEY POINTS

- Be sure to quote exactly.
- Distinguish between the two types of quotation marks.
- Use single quotation marks for quotations within quotations.
- Do not rely heavily on long, set-off quotations.
- Do not overuse quotation marks for emphasis.
- Remember the "inside" and "outside" rules when using other punctuation with quotation marks.

Correct use of quotation marks leads to clear, honest writing. Their misuse can cause misreading and, inadvertently, can make the writer guilty of plagiarism.

10.1 ALWAYS ENCLOSE SHORT DIRECT QUOTATIONS WITH QUOTATION MARKS.

Do not enclose material that is not part of the quotation in quotation marks.

EXAMPLES

"When I want results," the foreman said, "I always get less than I expect."

"We have waited too long to act," the president snapped. "Now we must pay the price."

10.2

Note: Some style manuals refer to "double" and "single" quotation marks. Normally, "double" quotation marks simply refer to normal quotation marks.

a. Use single quotation marks for quotations within quotations.

EXAMPLES

> The major said, "When the general shouts, 'Give me a report immediately!' he gets one."
> One of the technical editors remarked, "Although we allowed our writers to say 'It is us,' we do know formal grammar usage."

b. Double-space and indent ten spaces from the left-hand margin long quotations of more than four lines. Use quotation marks only if they appear in the original.

Avoid relying on blocks of quotations to make your statements. Use your own words.

EXAMPLE

At this point, one can still be quite certain of Candide's innocence:

> Our two travelers' first day was quite pleasant. They were encouraged by the idea of possessing more treasures than all Asia, Europe and Africa could collect. Candide in rapture carved the name of Cunegonde on the trees. (Voltaire, *Candide*, Random House, 1966, p. 151)

10.2 DO NOT USE QUOTATION MARKS WITH INDIRECT QUOTATIONS.

Weak	Improved
The president said that "Congress was not being fiscally responsible."	The president said that Congress was not being fiscally responsible.
My father told me that "the mayor was not very intelligent."	My father told me that the mayor was not very intelligent.

10.3

Note: Whether or not to use a direct quotation depends on how important the exact wording is. Of course, direct quotes help prevent misreading the speaker's or writer's message.

● **EXERCISE 1** In the following statements, insert or delete quotation marks wherever appropriate.

1. I am reminded, the general said, of the day Douglas MacArthur said to me, Let us not forget Bataan.
2. My supervisor said that "I was a very capable worker."
3. "There are too many of us in top management, the president snarled. I want some of you out on the line in the morning."
4. The book reviewer said that "Ivan Rassner's new book *Winning All the Way* should be read by everyone."
5. Thorndyke said, "I am sick of hearing Corrigan correct my writing by saying you can do better after he reads each page.
6. "Why are we all sitting here? our project engineer asked. Let's reformulate the problem."
7. I'm ready for the next voyage said Mr. Spock.
8. The DOD representative remembered I was working on the project when he was here before.

10.3 USE QUOTATION MARKS WITH THE TITLES OF SHORT WORKS SUCH AS SHORT STORIES, MAGAZINE ARTICLES, POEMS, AND SONG TITLES.

Note: Italicize titles of longer works. See Section 18.1, p. 116.

EXAMPLES

Gene Kelly's "Singing in the Rain" is the highlight of *Singing in the Rain.*
"Work expands to fill the time available" is a corollary of Parkinson's Law.

Note: When a short title is part of a longer title, underline (italicize) the longer title and enclose the shorter title in quotation marks.

10.5

EXAMPLES

> I borrowed *Critical Interpretations of "The Fall of the House of Usher"* to prepare for my exam.
>
> "Hands" is a story in *Winesburg, Ohio.*

10.4 FOR PURPOSES OF EMPHASIS OR IRONY, A WORD OR WORDS MAY BE ENCLOSED IN QUOTATION MARKS.

EXAMPLES

> The "kingdom" consisted of five acres of overgrown woodland on a dirt road twenty miles from town.
>
> When I look at baseball statistics, I want to see hits, runs, and runs batted in, not "on base percentage," "runs produced," and other statistical terminology.

10.5 DO NOT USE QUOTATION MARKS TO EMPHASIZE WORDS THAT DO NOT NEED EMPHASIS.

Weak	**Improved**
He believed that a "well-rounded education" was vital.	He believed that a well-rounded education was vital.
"Experience" really is the best teacher.	Experience really is the best teacher.

●**EXERCISE 2** Insert quotation marks wherever needed in the following sentences. Delete unnecessary ones.

1. I have always laughed when reading the book Invention Failures through the Ages.
2. He spent twenty years writing the essay Unified Field Theory or Why I Live at Belloien.
3. An example of his brilliance was his investing in a perpetual motion machine.
4. How many people realize that The Meaning of Relativity is a book by Albert Einstein?
5. His massive autobiography consisted of five pages mimeographed in faded purple ink.
6. The Fractal Geometry of Nature by Benoit Mandelbrot will always be one of my favorite books.

10.6

7. Dr. Franz told him he had a "complex personality."
8. The short story That Evening Sun contains the same characters as William Faulkner's novel *The Sound and the Fury.*
9. At any hour of the day or night, Meredith will sing Yes, We Have No Bananas.

10.6   USE OTHER PUNCTUATION MARKS CORRECTLY WITH QUOTATION MARKS.

a. Always place periods or commas inside quotation marks.

EXAMPLES

"Rasmus," the president said, "I want you to head up the new division."
"Possibly," the inventor hesitated, "someone in Arkansas had the same idea ten years ago."

b. Always place colons or semicolons outside the quotation marks.

EXAMPLES

The salesman said to me, "I'll give you a discount on a larger order"; unfortunately, I didn't need any more cartons at the time.
I told Hans grimly, "I have no time for such nonsense": Hans always insisted we obey minute rules.

c. Always place exclamation points and question marks within the quote marks when they apply *only* to the material being quoted; otherwise, place them outside the quotation marks.

EXAMPLES

Inside	Outside
The policeman asked, "Which way did he go?"	Why did the policeman say, "He went that way"?
Don't come in here and sing "I'll Be Glad When You're Dead, You Rascal You!"	I loathe it when you sing "Mary Had a Little Lamb"!

10.6

●**EXERCISE 3** Punctuate the following sentences correctly. If necessary, delete or change existent punctuation.

1. Bill Bailey said, "I'm coming home".
2. Ernest replied, "Do you know the way to Cawasee Corners"?
3. The physics professor said, "There are two kinds of theories;" he didn't tell us what they are.
4. The strongest language in a western movie of the 1940s was "gosh darn"!
5. He firmly believed in the old adage, "Never worry about the future".
6. Graham asked, "Why are computers so frustrating?"
7. "Good Lord"! the stranger said. "I thought this was St. Louis."
8. The chairman said, "I give you three reasons to live: stocks, bonds, and profiles."
9. "Because of you, we're finished", the chairman shouted at the vice president.
10. Whenever two or three people are together one of them will ask, "What shall we do now"?

●11

OTHER PUNCTUATION MARKS

KEY POINTS

- To prevent misreading, use all punctuation marks correctly.
- If uncertain of punctuation, read each sentence aloud.
- Do not overuse punctuation to create emphasis.

Other punctuation marks (the period, question mark, exclamation point, colon, dash, parentheses, brackets, slash, and ellipses) also must be used carefully. Although the use of these marks is not as complex as comma usage, without care you run the risk of being misread.

11.1 USE PERIODS AT THE END OF SENTENCES AND AFTER SOME ABBREVIATIONS.

a. Use a period to end declarative sentences or mild imperatives.

EXAMPLES

All students should study the rule book. [declarative]
Study the rule book. [mild imperative]
Everyone should bring a suggestion to the division meeting. [declarative]
Bring a suggestion to the division meeting. [mild imperative]

11.3

b. Use a period after some abbreviations: *Mr., Mrs., etc., et al.*

Note: Do not use periods with most abbreviations in ordinary writing, such as official postal abbreviations of state names and acronyms. The period is optional in the abbreviation *Ms.* If in doubt, consult a good dictionary for alternatives. See Section 13 for a complete discussion of abbreviations.

11.2 PUT A QUESTION MARK AFTER EVERY DIRECT QUESTION, BUT BE SURE TO DISTINGUISH BETWEEN DIRECT AND INDIRECT QUESTIONS.

EXAMPLES

What caused the drop in sales? [direct question]
The president asked what caused the drop in sales. [Indirect question]

a. Use a question mark with a question within a declarative sentence.

EXAMPLES

"What caused the drop in sales?" he asked.
"Who wants to balance the budget?" was my boss's favorite question every Friday afternoon.

b. Use a question mark within parentheses to indicate your uncertainty.

EXAMPLES

The Battle of Hastings was fought in 1023 (?) and affected English history tremendously.
Frederick Douglass was born in 1817 (?) and died in 1892.

11.3 USE THE EXCLAMATION POINT TO SHOW EMOTION, USUALLY AFTER AN INTERJECTION OR AN IMPERATIVE SENTENCE.

EXAMPLES

Stop that!   Hold your fire!
Ugh!   Wait until the machine stops!

11.4

Note: Do not rely heavily on the exclamation point. Save it for times when it is important to be emphatic. Use commas and periods with mild exclamations and imperative sentences.

EXAMPLES

Oh, I will be late.
How silently the time passes.
What a boring meeting this is.

Note: Do not use a comma or a period before or after an exclamation point.

● **EXERCISE 1** Punctuate the following sentences correctly.

1. My foreman asked me what made me come in late this morning?
2. Mr. Malone continually fell behind in his work.
3. "Where do you think you're going" my uncle asked.
4. Is this the way to Beetown.
5. There was some question as to why the cafe raised the price of their coffee?

11.4 USE A COLON TO INDICATE EXPLANATIONS OR SUMMARIES, FOR SALUTATIONS IN LETTERS, WHEN REFERRING TO BIBLICAL PASSAGES OR TO TIME, AND BETWEEN TITLES AND SUBTITLES.

a. Use a colon to indicate an explanation or summary.

EXAMPLES

He had three reasons to reject the bid: price, quality, and time.
I have found only one answer to obstinate ignorance: a quick departure.
She has two choices: a job as a technical writer or a job as a technical editor.

11.5

b. Use a colon after the salutation in a business letter (see Section 8 for further discussion).

EXAMPLES

Dear Professor Rodriguez:
Dear Mrs. Peterson:

c. Use a colon when referring to Biblical passages or to time, and between titles and subtitles.

EXAMPLES

Some people insist that the story of the Earth begins with Genesis 1:1.
Ivan signed in at 8:15 am.
Be sure to read *A Tycoon at Twenty: Memoirs of a Teen-Age Yuppie.*

d. Do not put a colon between a verb and its complement or object, or after *such as.*

Weak	**Improved**
The three companies merging were: US Steel, AT & T, and General Motors.	The three companies merging were US Steel, AT & T, and General Motors.
Many people prefer participatory sports, such as: rowing, swimming, and jogging.	Many people prefer participatory sports, such as rowing, swimming, and jogging.

11.5 USE THE DASH TO INDICATE A SUDDEN BREAK, TO SET OFF PARENTHETICAL ELEMENTS, AND AFTER AN INTRODUCTORY LIST OR SERIES.

Note: In typing, the dash is made with two hyphens. In handwriting, the dash is made with an unbroken line about the length of two typed hyphens.

a. Use the dash to indicate a sudden break in thought or change in tone.

EXAMPLES

I am reminded of my first job—but I digress.

I was able to find three candidates for the board presidency—in the nearest bar.

b. Use the dash to set off parenthetical elements—particularly for emphasis.

EXAMPLES

Edgar had three stages to his career—all bad.

Harold's credentials for the job were consistent—none were satisfactory.

Larry said he had the hottest job in town—he was the fire chief.

c. Use the dash after an introductory list or series.

EXAMPLES

Vain, pompous, skillful—the star had all of these qualities.

Farmer, lawyer, veteran—all three qualify for elective office.

Note: Do not overuse the dash. Especially do not replace commas, semicolons, or periods with dashes.

●**EXERCISE 2** Punctuate the following sentences correctly.

1. Ted is a fair manager for three reasons impartiality, sense of humor, and intelligence.
2. Three reasons Geraldine quit her job were: boredom, lack of support, and money.
3. He reported four times to his trusted ally his dog.
4. Happy, intelligent, hard-working, all McDougal Enterprises' employees seemed to possess these qualities.
5. I recently read *The Search for Moby Dick Under Sail for The White Whale.*
6. The CEO said that he owed his success in life to the one thing that had made him choose the cavalry, his childhood pony.

11.6 USE PARENTHESES TO SEPARATE PARENTHETICAL MATERIAL FROM THE REST OF THE SENTENCE.

EXAMPLES

> I was certain (or so I thought) that he would come.
> The manager wanted to see (1) if the contract was binding, or (2) if it could be revised.

Note: Commas and dashes also may be used to set off parenthetical material.

EXAMPLES

> There are many ways—according to Gomez—to gain control of a company.
> There are many ways, according to Gomez, to gain control of a company.

Note: Be sure to use parenthetical references to figures, documents, and parts of documents correctly.

EXAMPLES

> The line runs in at an 85-degree angle (see Figure 3).
> The intersection of Highways 151 and 81 has had too many accidents (refer to *Highway Accidents and Deaths, 1972–1986* for the complete statistics).

11.7 USE BRACKETS TO INTERPOLATE ADDITIONAL WORDS INTO A QUOTED PASSAGE OR TO REPLACE PARENTHESES WITHIN PARENTHESES.

EXAMPLES

> Doris said, "She [Anna Sage] is the best manager I've ever worked for."
> The notice read "All those wanting wurk [*sic*] report at dawn."
> There are a number of authorities (for example, Johnson on funding, Litwak on tax shelters, and [to a lesser degree] Harley on pension funds).

11.8   USE A SLASH TO INDICATE AN EITHER/OR SITUATION OR TO SEPARATE LINES OF POETRY.

a. Use a slash to indicate that either of two or more terms is acceptable.

EXAMPLES

Bring your own lawyers and/or accountants.
This is another sign that the time of a leaderless/aimless society has arrived.

b. Use a slash to separate lines of poetry.

EXAMPLES

Robert Frost's opening lines of "Mending Wall" are famous: "Something there is that doesn't love a wall / that sends the frozen ground swell under it."

11.9   USE ELLIPSES (THREE SPACED DOTS: . . .) TO INDICATE AN OMISSION FROM A QUOTED PASSAGE OR TO INDICATE HESITATION OR REFLECTION.

a. Use ellipses to indicate an omission from a quoted passage.

EXAMPLE

"We have three areas of concern: the lack of enthusiasm . . . the lack of funds . . . the lack of manpower. . . ." (Because this is the end of a sentence, a fourth period must be added.)

Note: Do not overuse ellipses with quoted material or the reader will suspect that more is being left out than is being included.

b. Use ellipses to indicate hesitation or reflection.

11.9

EXAMPLES

I had many reasons for declining the offer. . . . I still find them valid.
(Again, four dots are used because an entire sentence or sentences
have been omitted.)

"You're lucky I'm not going to say . . ." John blurted.

MECHANICS

●12

NUMBERS

- Numbers are important, so be sure they are accurate.
- Write numbers clearly, spelling out numbers below 10, at the beginning of a sentence, and when approximated.
- Clearly write all fractions and decimals using correct grammatical form.
- Use Roman numerals only when they are required.

Numbers express values that demonstrate a writer's knowledge about a subject. They often support specialized or highly technical material. Because of their importance in a text, they need to be accurate, without typographical mistakes, inconsistencies, or other errors.

12.1 SPELL OUT MOST NUMBERS BELOW 10.

In technical writing, spell out only small numbers. Even then, you will find many exceptions.

EXAMPLES

two days	five proposals
three wrenches	seven computers

12.2   WRITE AS FIGURES MOST OTHER NUMBERS.

a. Write the numbers 10 and above as figures.

Write figures for large numbers and those representing such items as measurable amounts, specific locations, or items of identification.

EXAMPLES

14 weeks	2,122 proposals
144 wrenches	17,000 computers

b. Write as figures all numbers expressing time, measurement, money, percentage, and proportion.

EXAMPLES

numbers expressing time:	7 pm, 1300 hours
date:	July 17, 1988 or 17 July 1989
measurement:	27 inches, 43 gallons
money:	700 dollars or $700
percentage:	8 percent or 8%
proportions:	3:2

c. Write as figures page numbers, identification numbers, highway numbers, and street addresses.

EXAMPLES

pages:	page 7, pages 99–102
identification:	Channel 9, Model 752e
highways:	Highway 151, Interstate 94 or I94
street address:	2339 Main Street

Note: One exception to the above rule occurs when the street address is *One.*

EXAMPLE

One University Plaza

12.3

12.3 WRITE AS FIGURES ALL NUMBERS IN TABLES AND FIGURES.

Tables and figures are meant to be easily read and to take up a limited amount of space.

EXAMPLE

TABLE 12.3a OXYGEN SOLUBILITY		
TEMPERATURE, °C	**OXYGEN, (MG/R)**	**% SATURATION**
1 0	10.0	68.4
2 5	8.1	63.3
3 10	6.3	55.6
4 15	6.9	68.0
5 20	5.8	63.2

●**EXERCISE 1** Where needed, supply the desirable style for the following sentences with numbers.

1. Jeffrey removed the dissecting scalpels by eight o'clock in the morning.
2. Acme Engineering paid 900 dollars for new software.
3. The rod measured seven inches in diameter and twelve inches in length.
4. A description of the extruder is on page three in the manual.
5. She once lived at 1 Langdon Street.
6. They traveled on Route Sixty Six to reach the desktop-publishing convention.
7. It took six days to achieve the correct angle.
8. When 7 compasses were stolen from his desk, Mr. Zarrow called a conference to determine what course of action should be taken.
9. Surefire Productions employs twenty-five people.
10. Highway 2 runs through Industrial Park.

12.4   WHEN SEVERAL NUMBERS APPEAR IN A SERIES, WRITE THEM ALL AS FIGURES.

To be consistent, write numbers in a series the same way each time a number appears.

Weak	**Improved**
John bought seven pencils, five pens, and 20 stencils.	John bought 7 pencils, 5 pens, and 20 stencils.

Note: When all the items in the listing are below ten, spell them out.

EXAMPLE

Jane made seven sales on Monday, nine on Wednesday, but only three on Friday.

12.5   SPELL OUT A NUMBER THAT BEGINS A SENTENCE.

Even if the number that begins a sentence would be normally written as a figure, spell it out. In cases where spelling out the number is awkward, rewrite the sentence because the number can be confusing to the reader.

Weak	**Improved**
20 days elapsed before he heard from the wholesaler.	Twenty days elapsed before he heard from the wholesaler.
11,781 workers enrolled in the pension plan.	In one month, 11,781 workers enrolled in the pension plan.
or	
Eleven thousand seven hundred and eighty one workers enrolled in the pension plan.	

●**EXERCISE 2**   Standardize the number usage in the following sentences.

1. Apex Machines purchased 25 computers, 1400 diskettes, and five new computer terminals.
2. 117 days elapsed before Jared heard from the company.

3. 1985 was an important growth year for Paulo Industries.
4. Seven months, 3 weeks, and 2 days after the warranty elapsed, the monitor stopped functioning.
5. 50 cases of thumbtacks had been ordered in error.

12.6 WHEN UNITS OF MEASURE FOLLOW NUMBERS, PLACE HYPHENS BETWEEN THEM IF THEY MODIFY A NOUN.

This rule applies to all numbers that are compounded with units of measure to form an adjective.

Note: *No* hyphen appears between the last modifier and the noun.

Weak	Improved
9 inch diameter hole	9-inch-diameter hole
10 year old typewriter	10-year-old typewriter
12 inch long cylinder	12-inch-long cylinder
300 yard dash	300-yard dash

●**EXERCISE 3** Correctly hyphenate the following units of measure.

1. The sewer line is 300 yards from the house.
2. Genevieve purchased a 6 pound portable computer.
3. The 75-inch tear made the polyethylene strip unusable.
4. From 240 feet the factory was impressive.
5. The crew was exhausted after a 36 hour layover in Kansas City.
6. Salty flew a 397 kilometer mission.
7. He refused to believe that the nylon tubing measured 6 inches in diameter.
8. A 3 mile hike every morning is invigorating.
9. In the center of the relay table was a 75 centimeter circle.
10. The company purchased 14 liters of glue.

12.7 WRITE FRACTIONS AND DECIMALS AS NUMBERS.

Fractions and decimals are difficult to read when they are spelled out.

Weak	Improved
three-fifths	3/5
seven-ninths	7/9
zero-point-two-five-four	0.254
nine-point-two-eight-three	9.283

12.8 PLACE A HYPHEN BETWEEN A WHOLE NUMBER AND A FRACTION WHEN TYPING MIXED NUMBERS.

For ease in reading typed copy, a hyphen separates the fraction from the whole number.

Weak	Improved
1 5/8	1-5/8
23 7/18	23-7/18
178 1/2	178-1/2

Note: Because typed fractions are more awkward to read than decimals, use decimals whenever possible.

12.9 LINE UP THE DECIMAL POINTS OF ALL NUMBERS CONTAINING DECIMALS.

EXAMPLES

0.715	2.1
93.24	8.63436
4.912	93244.4
0.5	

Note: Decimals below *1* require a *0* in front of them.

●**EXERCISE 4** Where necessary, revise the following fractions and decimals.

1. The pipe was one-half an inch too long.
2. The screw weighed zero-point-three lbs.
3. There was a depression in the aluminum sheeting every 2½ feet.
4. The weight was off by .2256 liters.
5. It is unusual for a formula to call for ⅔ of a gallon of solvent.

6. It took Jake 16⅞₁₆ seconds to find the spray gun.
7. He says he is right ninety-nine-point-two percent of the time.
8. John lined up the figures as follows:

$$0.275$$
$$0.26$$
$$321.04$$
$$75.328$$

12.10 IF A NUMBER IS ONE OR LESS THAN ONE, WRITE THE WORD IT MODIFIES IN THE SINGULAR.

Although it may seem proper to write "0.7 inches" instead of "0.7 inch," "0.7 inch" is grammatically correct. It helps to read the decimals as fractions spelled out.

Weak	Improved
0.723 tons	0.723 ton
1/6 inches	1/6 inch

Note: The one exception to this rule is the number zero. A zero is always treated as plural.

EXAMPLES

0 dollars
0 units

12.11 SPELL OUT NUMBERS THAT ARE APPROXIMATIONS.

Approximations are averages, guesses, or amounts close to the actual figure. If you write them as numbers, you imply they are precise and therefore mislead the reader.

EXAMPLES

three quarters of a mile	about six inches
half an acre	two thirds completed

Note: If a whole number such as the "six inches" given in the previous example is an estimated value, be sure to preface the amount with a qualifier such as "about" or "approximately," or your reader will not know it is an approximation.

Do not approximate a mixed number. If a number is that close, give the exact figure.

Weak	Improved
approximately nine-point-three feet	approximately nine feet or 9-⅓ feet
approximately nine and three-tenths feet	approximately nine feet

12.12   SPELL OUT THE SMALLER OF TWO NUMBERS THAT APPEAR TOGETHER.

Weak	Improved
250 7-watt light bulbs	250 seven-watt light bulbs
16 9-member softball teams	16 nine-member softball teams
11 5,000 BTU air conditioners	eleven 5,000-BTU air conditioners
7 1988 marketing reports	seven 1988 marketing reports
56 6th Street	56 Sixth Street

●**EXERCISE 5**   Revise any numbers that need it.

1. The milling machines weighed about 500 pounds.
2. Suzanne ordered 75 2-sided, double-density diskettes.
3. The factory covers approximately an acre.
4. Paula raised the unit to ½ inches.
5. Luis reported his work on the Damson project was ⅔ completed.

12.13   WRITE LONG NUMBERS CLEARLY.

Not all sources agree on how to use the several ways to write long numbers. In general, write numbers from 1,000 to 999,999 as follows:

1,000	100,000
10,000	999,999

Write round numbers in the millions and more in three ways: as figures, as a figure with the rest spelled out, or as a scientific notation.

EXAMPLES

as figures:	1,000,000
	1,250,000
	1,000,000,000
as part-figure:	1 million
	1.25 million
	1 billion
as a scientific notation:	$1 * 10^6$
	$1.25 * 10^4$
	$1 * 10^9$

Note: Use numbers if room is available; use scientific notation only when space is at a premium. If you cannot round off a number, write it out.

EXAMPLES

11,671,832
123,685,101

12.14 USE ROMAN NUMERALS ONLY IN SPECIAL CASES.

a. Use lowercase Roman numerals to number the front material of a formal document.

Such material as the table of contents, abstract or summary, glossary of terms, symbol charts, acknowledgments, and the like are placed in the front of a formal report (see Section 36 for a discussion of the formal report). These pages are numbered with lowercase Roman numerals: i, ii, iii, iv, v, vi, and so on.

b. Use Roman numerals for the main headings in a formal outline.

The formal outline is discussed in detail in Section 19.6, p. 122.

c. Use Roman numerals after the names of monarchs, popes, and sons from the second generation on with the same names.

EXAMPLES

Louis XIV	Elizabeth II
Pope Pius X	Reginald Vanderbilt III

d. Use Roman numerals to show passages from plays.

EXAMPLE

The passage was from Shakespeare's *King Henry the Sixth*, II, iii. (The *II, iii* refers to the third scene in the second act of the play.)

TABLE 12-14 FORMATION OF ROMAN NUMERALS

Arabic Numerals	Roman Numerals	Arabic Numerals	Roman Numerals
1	I	50	L
2	II	60	LX
3	III	70	LXX
4	IV	80	LXXX
5	V	90	XC
6	VI	100	C
7	VII	110	CX
8	VIII	199	CXCIX
9	IX	200	CC
10	X	400	CD
11	XI	499	CDXCIX
15	XV	500	D
19	XIX	900	CM
20	XX	999	CMXCIX
21	XXI	1000	M
29	XXIX	1500	MD
40	XL	3000	MMM

12.14

●**EXERCISE 6**   Write the following numbers as directed.

1. Turn 3,000,000,000 into a scientific notation.
2. Write out the following scientific notation: $7.5 \cdot 10^8$.
3. Write 3,500,000 spelled out in words.
4. Turn 3 trillion into a number.
5. Turn the following Arabic numerals into Roman numerals: 875; 67; 1,234; 484; 3,299.

●13

ABBREVIATIONS

KEY POINTS

- Too few is better than too many abbreviations.
- Be sure the intended reader will understand what an abbreviation stands for.
- Clearly define abbreviations that the reader might not understand.

Abbreviations are shortened forms of words. They save time in writing, making documents shorter to produce. Unfortunately, documents tend to suffer from too many rather than too few abbreviations. Often these abbreviations are peculiar to a business group. Once a document leaves this group, its abbreviations can create confusion for outsiders. The best rule for any writer is to avoid abbreviations whenever possible unless the intended audience is clearly known and will understand the abbreviations. When this is not possible, abbreviations should be clearly defined and used consistently. In general, abbreviations are confined to graphics, where space is minimal.

13.1 USE ABBREVIATIONS FOR TITLES ONLY WHEN THE ABBREVIATIONS ARE APPROPRIATE.

a. Use abbreviations that are commonly known and employed.

Use generally understood abbreviations, such as *Mr.*, *Mrs.*, or *Dr.*, but avoid abbreviated titles that are understood only by a few.

Weak	Improved
When he retired, the EE moved to Florida.	When he retired, the electrical engineer moved to Florida.
The CP lives next door to me.	The computer programmer lives next door to me.
Ad. Dir. Larsen can be very demanding at times.	Administrative Director Larsen can be very demanding at times.

b. Do not begin and end a name with the same title.

This practice is redundant, therefore unnecessary.

Weak	Improved
Dr. John Gutzmacher, DDS	Dr. John Gutzmacher *or* John Gutzmacher, DDS
Atty. Janice Quinn, LLB	Atty. Janice Quinn *or* Janice Quinn, LLB
Dr. Arnold Anderson, PhD	Dr. Arnold Anderson *or* Arnold Anderson, PhD

Note: If a person has more than one title, include all of them *after* the name.

EXAMPLES

Carl Carson, PhD, PE
Roxanne Gomez, PhD, MD

c. Do not abbreviate titles of respect or of great importance.

Words like the following show respect or importance:

Honorable	President
Reverend	Governor

d. Do not abbreviate people's names.

When you abbreviate a person's name, you show lack of respect. In addition, the abbreviation can be confusing if the reader is not familiar with it.

13.2

EXAMPLES

Abbreviation	Correct Name
Chas.	Charles
Eliz.	Elizabeth
Geo.	George
Jas.	James

e. Abbreviate academic, religious, military, and civilian titles only when they are followed by a person's name.

Weak	Improved
The dr. examined Jake.	The doctor examined Jake.
Doctor Lewis examined Jake.	Dr. Lewis examined Jake.

13.2 ABBREVIATE COMPANY NAMES ONLY IF THE COMPANY DOES.

IBM, GTE, and NCR are acceptable abbreviations because the companies use them, but *Bro.* for Brother or *BC Co.* for Barber Coleman Company are not acceptable because the companies do not employ these abbreviations.

●**EXERCISE 1**   Change each of the following to an acceptable form.

1. The Hon. Justice Loto presided over the court.
2. Dr. John Sampson, M.D., gave the after-dinner speech.
3. The letter was addressed to Robt. McConnell.
4. Rollo Stein had always wanted to work for International Business Machines.
5. The Rev. Jackson gave the invocation.
6. Prof. Allan's lecture on nutrients held her class's interest.
7. Eliz. and Thos. Strang were promoted after they had more sales last month than anyone else in the department.
8. Mister Parker and Doctor Chambers attended the executive meeting.
9. Gov. Gorganzo lost the election after a bitter, expensive campaign.
10. Royce was promoted to assist. mgr. of his dept.

13.3 OMIT INTERNAL PUNCTUATION IN MOST ABBREVIATIONS.

The practice of using internal punctuation in abbreviations is rapidly dying out. Use it only in company names where a company still uses it, such as M.A.P. Systems, Inc. Internal punctuation is unnecessary for meaning and is time-wasting to type. In addition, omitting internal punctuation is neater, cleaner, and more readable.

Weak	Improved
7 A.M. or a.m.	7 AM or am
U.S.A.	USA
U.S.S.R.	USSR
I.B.M.	IBM
A.T.&T.	AT&T
m.p.h.	mph
r.p.m.	rpm

13.4 USE TERMINAL PUNCTUATION WHEN THE ABBREVIATION ENDS IN A LOWERCASE LETTER.

EXAMPLES

Titles:	Dr., Mr., Mrs.
Addresses:	Ave., St., Blvd., Rd.
References:	p., pp., Ch., Fig., no.
Businesses:	Co., Inc., Corp., Ltd.
Days, months:	Mon., Sat., Jan., Sept.

Note: The optional *Ms* is an exception to the rule.

Note: When using postal abbreviations for states, do not use a period. See Section 13.7, p. 96, for a list of these abbreviations.

●**EXERCISE 2** Correct any of the following abbreviations that are inconsistent with the rules.

1. Mr Jones is the first supervisor to accept the metric system at our plant.
2. We sent the shipment to 321 Market St, but it never arrived.
3. After working for the French branch of the company for six years, the Altendorfs were happy to return to the U.S.A.

13.6

4. The man was so old he could remember belonging to the CIO.
5. Marcia bought a GTE telephone.
6. The equation was on p. 3.
7. Morris has been transferred to the 6 a.m. shift.
8. Allen asked for a no 7-gauge wire.
9. Jane moved her office to Third St South.
10. After he graduated from college, he got a job with the C.I.A.

13.5 WRITE OUT WORDS TO BE ABBREVIATED THE FIRST TIME THEY ARE USED, THEN INCLUDE THEIR ABBREVIATED FORM IMMEDIATELY AFTERWARD IN PARENTHESES.

EXAMPLES

The computer assisted design (CAD) software has been improved. CAD is now used extensively in many companies' operations.

The Manufacturers' Activity Board (MAB) is meeting this afternoon to vote on how it will spend the $5,000 collected in rental fees. The MAB regrets that so few employees take an interest in what it does.

13.6 BE SURE THE READER UNDERSTANDS AN ACRONYM OR OTHER RELATED ABBREVIATIONS BEFORE YOU EMPLOY THEM.

Acronyms and related abbreviations are titles formed from the initial letters of the words that make it up, such as NATO, NASA, and UNICEF, which are fairly well-known. Uncommon titles should be written as words the first time for lay audiences.

Acronyms and related abbreviations are used frequently in scientific and technical writing, where long phrases are needed repeatedly. Some lesser-known yet common acronyms and abbreviations from several technical fields are listed below. For an extensive list of acronyms and related abbreviations in many technical fields, see Emanuel Benjamin Ocran, *Ocran's Acronyms* (London: Routledge & Kegan Paul, 1978).

Aeronautics

AAA Anti Aircraft Artillery
ARB Air Registration Board
CAA Civil Aeronautics Administration

ETD	Estimated Time of Departure
ETA	Estimated Time of Arrival
FAA	Federal Aviation Agency
MSL	Mean Sea Level
NAM	Nautical Air Miles
UFO	Unidentified Flying Object

Agricultural Sciences

AI	Artificial Insemination
CP	Crude Protein
FDA	Food and Drug Administration
LNPK	Limestone Nitrogen Phosphorus and Potassium
PCA	Production Credit Association

Chemical and Biological Sciences

AAS	Atomic Absorption Spectroscopy
BOD	Biological Oxygen Demand
DDT	Dichlorodiphenyl Trichloroethane
DNA	Deoxyribonucleic Acid
DO	Dissolved Oxygen
EDTA	Ethylenedramine Tetra Acetic Acid
GC	Gas Chromatograph
GC-MS	Gas Chromatography-Mass Spectrometry
HDL	High Density Lipoprotein
HPLC	High Pressure Liquid Chromatography
LD	Lethal Dose
PAC	Polycyclic Aromatic Hydrocarbons
PCB	Polychlorinated Biphenyl
THM	Trihalomethane
TLC	Thin Layer Chromatography
TOC	Total Organic Carbon

Computer Sciences

AD	Analog-to-Digital
ASCII	American National Standard Code for Information Interchange
BASIC	Beginner's All-purpose Symbolic Instruction Code
CAD/CAM	Computer Aided Design/Computer Aided Manufacturing
COBOL	Common Business Oriented Language
CPU	Computer Processing Unit

DOS   Disk Operating System
EBCDIC   Extended Binary Coded Decimal Interchange Code
EOF   End of File
KB   Kilobytes
RAM   Random Access Memory
ROM   Read Only Memory

Engineering

ABET   Accreditation Board for Engineering and Technology
ASCE   American Society of Civil Engineers
ASME   American Society of Mechanical Engineers
ASTM   American Society for Testing of Materials
CEC   Consulting Engineers Council
IEEE   International Electrical & Electronics Engineers
MSL   Mean Sea Level
OSHA   Occupational Safety and Health Administration
PE   Professional Engineer
RC   Reinforced Concrete
TDS   Total Dissolved Solids

Space Sciences

EROS   Earth Resources Observation Satellite
JPL   Jet Propulsion Lab
NASA   National Aeronautics and Space Administration
NOAA   National Oceanographic and Atmospheric Administration

Transportation

AASHO American Association of State Highway Officials
BART   Bay Area Rapid Transportation
CTA   Chicago Transit Authority
DOT   Department of Transportation
CHP   California Highway Patrol
SAE   Society of Automotive Engineers

Note: Technically an acronym must be a pronounceable word; however, other abbreviations, such as FBI, CIA, and NFL, are usually referred to as acronyms. As a result, a more accurate definition of an acronym would be "all capitalized abbreviations that do not employ periods between the letters."

13.8

13.7 WHEN ABBREVIATING THE NAMES OF STATES, TERRITORIES, AND POSSESSIONS, USE THE UNITED STATES POSTAL ABBREVIATIONS.

The correct abbreviations are given in the list below.

AL	Alabama	**KY**	Kentucky	**ND**	North Dakota
AK	Alaska	**LA**	Louisiana	**OH**	Ohio
AS	American Samoa	**ME**	Maine	**OK**	Oklahoma
AZ	Arizona	**MD**	Maryland	**OR**	Oregon
AR	Arkansas	**MA**	Massachusetts	**PA**	Pennsylvania
CA	California	**MI**	Michigan	**PR**	Puerto Rico
CZ	Canal Zone	**MN**	Minnesota	**RI**	Rhode Island
CO	Colorado	**MS**	Mississippi	**SC**	South Carolina
CT	Connecticut	**MO**	Missouri	**SD**	South Dakota
DE	Delaware	**MT**	Montana	**TN**	Tennessee
DC	District of Columbia	**NE**	Nebraska	**TX**	Texas
FL	Florida	**NV**	Nevada	**UT**	Utah
GA	Georgia	**NH**	New Hampshire	**VT**	Vermont
GU	Guam	**NJ**	New Jersey	**VI**	Virgin Islands
HI	Hawaii	**NM**	New Mexico	**WA**	Washington
ID	Idaho	**NY**	New York	**WV**	West Virginia
IL	Illinois	**NC**	North Carolina	**WI**	Wisconsin
IN	Indiana			**WY**	Wyoming
IA	Iowa				
KS	Kansas				

13.8 DO NOT CONFUSE THE LATIN ABBREVIATIONS *e.g.* AND *i.e.*

The first, *e.g.*, is the abbreviation for *exempli gratia*, which means "for example." The abbreviation *i.e.* stands for *id est*, or "that is." Use *e.g.* with examples and *i.e.* for clarification.

EXAMPLES

The supervisor told everyone to "shut down," i.e., he wanted his crew to be quiet. (The abbreviation *i.e.* is used to clarify what the supervisor said.)

We ordered a number of odds and ends for the office: e.g., paper clips, rubber bands, blue pencils, and ballpoint pens. (The abbreviation *e.g.* is used to give examples of the items ordered.)

13.8

Note: The tendency today is *not* to underline familiar Latin abbreviations. See Section 18.2, p. 118.

● **EXERCISE 3**   Supply appropriate acronyms or abbreviations for each of the following.

1. She belongs to the Business and Professional Women's Association.
2. His address is 901 Main Street, Fish Springs, Alaska.
3. Joe once played for the American Football League.
4. He fought in both World War One and World War Two.
5. Sarah graduated from Southern Methodist University in 1985.
6. Harcourt Brace Jovanovich is one of the major publishers of college textbooks.
7. The company was located at 9211 Industrial Park Road, Rockford, Illinois 61108–0004.
8. When Ursula was in New York, she visited Wall Street.
9. Daughters of the American Revolution is a very old, established organization in the U.S.A.
10. She sent the letter to 671 Clarksling Street, Preston, Maryland 21655.

●14

UNITS OF MEASURE

KEY POINTS

- Write out units of measure whenever suitable.
- Use units of measure consistently and correctly.

Often a piece of writing requires many references to units of measure. Consider your audience's needs when deciding how to use them.

14.1 IN GENERAL, USE TYPICAL SYMBOLS RATHER THAN WRITING OUT THE WHOLE WORD FOR UNITS OF MEASURE.

Technical people are often too busy to write out each unit of measure; therefore, it is wise to use standardized symbols. For the distinction between technical-writing symbols and abbreviations, see Section 15.

EXAMPLES

Term	Acceptable	Unacceptable
ampere	A	amp
kilogram	Kg	kilo
Kelvin	K	Kel

14.2   USE ALL UNITS OF MEASURE CONSISTENTLY.

Do not mix English and metric systems. Choose a set of units and use it consistently throughout the text and the graphics.

Weak	Improved
The cylinder was 15 cm in diameter and 12 in long.	The cylinder was 15 cm in diameter and 30 cm long.
	or
	The cylinder was 6 in. in diameter and 12 in long.

14.3   REFER TO SECONDARY UNITS IN PARENTHESES AFTER THE PRIMARY UNITS.

It is not always necessary to include two systems of measurement, but when it is, select one system as secondary and place it in parentheses after the primary one.

EXAMPLES

The title contained 28 grooves per 10 cm (4 in.).
The company ordered 70 meters (76.55 yards) of wire.

14.4   EMPLOY ABBREVIATIONS WITH UNITS OF MEASURE ONLY WHEN THEY ARE PRECEDED BY OR FOLLOWED BY A NUMBER.

EXAMPLES

75°F   12V
9 ft 7 in   44 cm

Note: This same rule holds true with other abbreviated words. For example, "222 pp" is correct, but in the following sentence, this abbreviation would be incorrect: "The pp are falling out of my calculus book." Another example, "See Fig. 1," is acceptable, as is "The figures are well-drawn."

14.5

14.5 NEVER USE A PLURAL ENDING WITH ABBREVIATIONS FOR UNITS OF MEASURE, EVEN WHEN THE UNITS ARE USED IN THE PLURAL.

Weak	Improved
22 lbs	22 lb
17 ins	17 in
55 cms	55 cm

● **EXERCISE 1**   Revise the following units of measure.

1. The iron bar was 12 cm in diameter and 12 in long.
2. The company ordered many in. of wire.
3. The cartridge weighed only 12 lbs.
4. Changing °F to °C can be confusing.
5. Sylvia Loraski ordered 8 in.—20 cm—of tape.
6. Last year Rita Gonzalez won our company picnic's 300-yds-dash.
7. The thermometer registered 5° Celsius.
8. After 12 pounds were added to the aircraft engine, even an extra gram would have been too heavy.
9. The chlorine is bottled in ltrs.
10. Yesterday it was 94° F in the shade.

●15

SYMBOLS

KEY POINTS

- Carefully define symbols.
- Be sure symbols are not duplicated.
- Punctuate symbols correctly.
- Use correct symbols, *i.e.*, symbols that are widely accepted.

A symbol is usually defined as something that stands for something else. In technical writing, symbols are letters or signs that designate a mathematical term or a unit of measure. Symbols are to be used with caution, for not everyone outside a particular technical field understands them. Like abbreviations, they do save typing time as well as space and are often confined to graphics, where space is at a premium.

15.1 DEFINE ALL SYMBOLS CAREFULLY.

Readers must understand the symbols used. Do not send your readers to technical references to find out a symbol's meaning. Rather, define the symbols used in the text and place a list of nomenclatures (definitions) at the beginning or end of the document to clarify all symbols used.

EXAMPLE

LIST OF SYMBOLS USED IN THIS DOCUMENT

SYMBOL	DEFINITION	SYMBOL	DEFINITION
Q	Flowrate	C	Concentration
μ	Absolute Viscosity	g	Gravitational Acceleration
K_H	Henry's Law Constant	γ	Specific Weight
Re	Reynolds No.	ρ	Density

15.2 DO NOT DUPLICATE SYMBOLS.

A symbol may stand for more than one unit of measure or variable. Be sure to use symbols that are not duplicated. If they are, write out all the terms as words or use carefully defined new symbols. For example, K can stand for Kelvin as well as for kilobytes, potassium, and reaction rate. If more than one of these terms appear in the same document, use a new symbol for one of the terms after carefully defining the new symbol for the reader.

15.3 OMIT PUNCTUATION IN SYMBOLS.

The only exception to this rule arises when the symbol spells a word, as with the symbol for inches, *in.* Use *in.* in complete sentences where *inch* (or *in*) is followed by the preposition *in.* Use *in* for graphics and in complete sentences where it is not followed by the preposition *in.*

SENTENCE EXAMPLES

The cylinder was 22 in. in diameter.
He placed the 22-in-diameter cylinder on the loader.

TABLE EXAMPLE

TABLE 15.3A METRIC CONVERSION FACTORS FOR LENGTH

US NAME	SYM-BOL	MULTI-PLY BY	TO OBTAIN	SI NAME	SYM-BOL
foot	ft	0.3048		meter	m
inch	in	2.54		centimeter	cm
inch	in	0.0254		meter	m
inch	in	25.4		millimeter	mm
mile	mi	1.6093		kilometer	km
yard	yd	0.9144		meter	m

15.4 USE CORRECT GRAMMAR WHEN USING SYMBOLS AS WORD SUBSTITUTES.

Weak	**Improved**
In the equation $s = x^2 + 2$, x are the variables.	In the equation $s = x^2 + 2$, x is the variable.

15.5 DO NOT USE ' FOR FEET OR " FOR INCHES.

The symbols ' and " are used in geography to express minutes and seconds of latitude and longitude.

EXAMPLE

4°10′12″ S

Note: The one exception to this rule is in blueprints, where ' for feet and " for inches are necessary.

15.6 DO NOT USE *X* WITH DIMENSIONS IN WRITTEN PROSE.

It is common practice to write dimensions as follows: 3″ × 7′ × 6″. However, using symbols in this manner should be confined to blueprints and to scratch notes. In written documents, the above dimensions should be written as follows:

The board measured 3 in by 6 in by 7 ft.

15.7 AVOID UNNECESSARY TYPEWRITER SYMBOLS OR SIGNS.

The use of @, **#**, and **&** can be confusing for the reader.

Confusing	Clearer
He sold six reams @ $5 each.	He sold six reams at $5 (or 5 dollars) each.
He needed #2 lead for his pencil.	He needed number 2 lead for his pencil.
The logo was red, white, & green.	The logo was red, white, and green.

15.8 WHEN USING SYMBOLS FOR EITHER THE US SYSTEM OR THE METRIC OR INTERNATIONAL SYSTEM OF UNITS (SI), BE SURE TO USE THE CORRECT SYMBOL.

The table below lists a number of the symbols used in the two systems.

TABLE 15-8 US AND METRIC TERMS AND SYMBOLS

US Term	Symbol	SI Term	Symbol
Area			
square inch	in^2	square centimeter	cm^2
square foot	ft^2	square meter	m^2
square mile	mi^2	square kilometer	km^2
acre	A	hectare	ha
Length			
inch	in	millimeter	mm
foot	ft	centimeter	cm
yard	yd	meter	m
mile	mi	kilometer	km

US Term	Symbol	SI Term	Symbol
Mass			
ounce	oz	gram	gm
pound	lb	kilogram	kg
ton	t	megagram	mg
Temperature			
degrees Fahrenheit	°F	degrees Celsius (centigrade)	°C
		degrees Kelvin	°K
Volume			
cubic inch	in³	cubic centimeter	cm³
cubic foot	ft³	cubic meter	m³
cubic yard	yd³	liter	L
acre-ft	acre-ft		
ounce	oz		
gallon	gal		

● **EXERCISE 1**   Change any symbols that are not used correctly.

1. The new office measured $75'6'' \times 120'16''$.
2. Everyone knows that $a^2 + b^2 = c^2$ are a well-known equation.
3. The desk was 5 ft. by 2 ft. by 30 in. high.
4. The liquid was channelled for 55 yds before gas was induced into it.
5. Even .0256 gram of the powder could be lethal.
6. Charles decided to use the symbol A to represent area.
7. The rod was 7 in in diameter.
8. The blueprint showed the distance between the window and the door was $3'8''$.
9. When Joe saw the symbol K, he assumed it referred to *Kelvin*.
10. Even 15 cm. is too great a length for the rod.

●16

EQUATIONS

- Always display equations as clearly as possible in the text.
- Proofread each equation for accuracy.
- If software or typewriter keys are not available for symbols, use press-on stickers or print the symbols neatly in black ink.

Some equations are simple to incorporate in a typed text. For example, $a^2 + b^2 = c^2$ can be typed easily into a text. On the other hand, even a short equation like $A = \pi r^2$ demonstrates the main problem with equation writing: symbols must be put in by hand. Software and special typewriter balls or wheels are available for some symbols, but in many cases the writer must add these symbols manually.

16.1 CENTER LONG EQUATIONS AND PARENTHETICALLY NUMBER THEM IN THE TEXT; INCLUDE SHORT EQUATIONS AS PART OF THE WRITTEN TEXT.

DISPLAYING A SHORT EQUATION

To find the circumference of a circle, use the equation $C + \pi d$.

16.3

DISPLAYING A LONG EQUATION

$$h = \frac{fL(1 - e)V_s^2}{e^3 g d_p} \tag{1}$$

Note: Long equations are always numbered to simplify references to them later in the text.

EXAMPLE

To calculate the head loss, see Equation 1.

16.2 WRITE EQUATIONS SO THAT ALL FRACTIONS, DIVISION LINES, AND PLUS AND MINUS SIGNS ARE ALIGNED AND CLEAR IN MEANING.

Incorrect Meaning	Correct Meaning
$y - k = a/b(x - h)$	$y - k = (a/b) \cdot (x - h)$

16.3 IF IT BECOMES NECESSARY TO BREAK A LONG EQUATION, DO SO AT AN EQUAL SIGN OR AT A PLUS OR MINUS SIGN THAT IS NOT IN PARENTHESES OR BRACKETS.

BREAKING AT AN EQUAL SIGN

$$a_0\left(x^2 + \frac{a_3}{a_0}x + \frac{a_3}{2a}\right)^2 + a_2\left[y^2 + \frac{a_4}{a_2}y + \left(\frac{a_4}{2a_2}\right)^2\right]$$
$$= -a_5 + \frac{a_3^2}{4a_0} + \frac{a_4^2}{4a_2} \tag{1}$$

BREAKING AT A PLUS OR MINUS SIGN

$$y = 113.1528x^2 + 103.4712x^4 + 333.8927x^3 + 22138.8764x^2$$
$$+ 2222.1938x + 2008.6935$$

16.4 IF EQUATIONS DIRECTLY FOLLOW ONE ANOTHER IN A WRITTEN TEXT, PRESENT THEM ON SEPARATE LINES, REGARDLESS OF THEIR LENGTH.

Weak

If $L_e = \dfrac{L(1 - e)}{1 - e_e}$ and $e_e = \left(\dfrac{V}{v}\right)^{.22}$, then $L_e = \dfrac{L(1 - e)}{1 - \left(\dfrac{V}{v}\right)^{.22}}$.

Improved

If

$$L_e = \frac{L(1 - e)}{1 - e_e}$$

and

$$e_e = \left(\frac{V}{v}\right)^{.22}$$

then

$$L_e = \frac{L(1 - e)}{1 - \left(\frac{V}{v}\right)^{.22}}$$

16.5 PUNCTUATE EQUATIONS IN THE SAME WAY AS ANY OTHER WRITING.

a. Use no punctuation when introducing an equation with a simple (subject + predicate) sentence.

EXAMPLE

Some equations are identities. An example is

$$x + 1 = x + 1$$

b. When introducing an equation with an introductory phrase or clause, place a comma after the phrase or clause.

EXAMPLE

As stated before,

$$x + 1 = x + 1$$

is an *identity*.

c. Use a colon when introducing an equation at the end of a complete sentence.

EXAMPLE

The following equation is an example of an identity:

$$x + 1 = x + 1$$

16.6   INDICATE MULTIPLICATION BY AN ASTERISK (*) AND DIVISION BY A SLASH (/).

Weak	Improved
$y - k = (a \div b) \times (x - h)$	$y - k = (a/b) * (x - h)$

16.7   DISPLAY ALL EQUATIONS IN A PROFESSIONAL MANNER.

In a rough draft, handwrite any equations that contain symbols. For final copy, if computer software or typewriter keys for symbols are not available, either use press-on stickers for symbols or write the symbols neatly in black ink. Be sure to proofread all equations carefully, no matter who types them.

●17

CAPITALIZATION

KEY POINTS

- Capitalize proper names and their derivations.
- Capitalize titles.
- Use capitalization judiciously. If in doubt, consult a recent college or desk dictionary for advice.

Most capitalized words fall into one of six classes: proper names, trade names, key words in titles, derivations of proper names, and first words in sentences or dialogue. Through the years, however, capitalization has undergone some changes. For example, *Frisbee* is generally written *frisbee*, *Aspirin* is usually *aspirin*, and *ping-pong* is rarely written as *Ping-Pong*, its correct trade name. Other words are sometimes but not always capitalized. For example, Janet went out *West* last summer, but back home she lives on the *west* side of town. She also lives on the planet *Earth*, but the *earth* around her house is mostly clay.

Because of these and other usages, it is prudent to consult a current college dictionary if in doubt about the capitalization of a word.

17.1 CAPITALIZE PROPER NOUNS.

a. Capitalize proper names of persons, places, or things.

17.1

EXAMPLES

Anna Gustafson	Chicago Bears
University of Virginia	the Declaration of Independence
Memorial Day	the Bible
San Diego	South Pacific

Note: Do *not* capitalize classes of persons, places, or things.

Capitalized	Not Capitalized
Sir Arthur Conan Doyle	the author
The University of Arizona	a university
Thanksgiving Day	a day of thanksgiving
Easter	spring vacation
Washington, D.C.	the nation's capital
Chicago Cubs	a major-league baseball team
the Bill of Rights	a document

Note: When referring to the President of a country, *President* is sometimes capitalized.

b. Capitalize trade names.

EXAMPLES

Xerox	Hotpoint
Kleenex	Coca Cola

c. Capitalize words used as derivatives of proper nouns.

EXAMPLES

Illinoisan	Leninite
Martian	Wesleyan

d. Capitalize government agencies, departments, and divisions, and organizations and companies.

EXAMPLES

government agencies:	Justice Department
departments:	Department of Natural Resources

111

divisions:	Division of Motor Vehicles
organizations:	Society for Technical Communications
companies:	General Electric Corporation

e. Capitalize titles preceding names.

EXAMPLES

Private First Class Thomas Ward   President Patricia Jasper
Judge Bruce Talso   Professor Esther Burke

f. Capitalize names of religions, holy days, and all words that mean the Supreme Being.

EXAMPLES

Christianity	All Saints Day	God
Judaism	Good Friday	Jehovah
Islam	Yom Kippur	Lord
Buddhism	Channukah	Allah

Note: Do not capitalize adjectives derived from religious terms.

EXAMPLES

biblical	godlike
saintly	buddhist monk

g. Capitalize days of the week, months, and holidays.

EXAMPLES

Monday	Christmas
November	Fourth of July

● **EXERCISE 1** Supply capitals where needed.

1. Second city is a well-known name for Chicago.
2. Three years ago professor Swartz accepted a position at the Illinois institute of technology.
3. The company was closed for labor day weekend.
4. Last week three japanese executives toured the plant.

5. The civil engineer accepted a job with the michigan department of public transportation.
6. Every memorial day our department has a picnic.
7. On monday night our project engineer wants our progress reports.
8. The demonstration by the buddist priests held up traffic.

17.2   CAPITALIZE THE FIRST WORD IN A SENTENCE AND THE FIRST WORD IN A DIRECTLY QUOTED SPEECH.

In the following example, the first word in each sentence is capitalized. The first word in the dialogue also is capitalized because it is a direct quotation of speech.

Capitalized	Not capitalized
Alex said, "They attended an educational conference."	Alex said they attended an educational conference.

Note: Do not capitalize directly quoted speech if it does not begin a sentence.

EXAMPLE

Alex said he attended "the most educational conference the company has ever sponsored."

17.3   CAPITALIZE THE FIRST LETTER OF ALL WORDS IN TITLES, EXCEPT ARTICLES (A, AN, THE), COORDINATING CONJUNCTIONS (AND, BUT, OR, YET, SO, FOR, NOR), AND PREPOSITIONS.

The above rule is true of all titles—books, plays, journal articles, research papers, reports, and the like.

Note: One exception—capitalize articles, prepositions, and conjunctions when they begin a title.

EXAMPLES

The Handbook of Technical Writing [article, *the*]
To Each His Own [short preposition, *to*]
Yet Shall He Know Them [coordinating conjunction, *yet*]

17.4 DO NOT CAPITALIZE WORDS TO EMPHASIZE THEIR IMPORTANCE.

Underline (italicize) emphasized words.

Weak	Improved
Joan said it was IMPERATIVE that we attend the meeting.	Joan said it was *imperative* that we attend the meeting.

17.5 CAPITALIZE NAMES OF COUNTRIES, NATIONALITIES, AND LANGUAGES.

EXAMPLES

Switzerland	English
Rhodesian	Greek

17.6 CAPITALIZE CERTAIN SINGLE LETTERS.

Capitalize the pronoun *I*, the interjection *O*, letters used with hyphens to begin individual words, such as *X-ray* and *T-square*, and letters that serve as words, such as *grade A*.

17.7 CAPITALIZE ABBREVIATIONS AND ACRONYMS THAT ARE SHORTENED FORMS OF CAPITALIZED WORDS.

EXAMPLES

NYC (New York City)
CNN (Cable Network News)
HBO (Home Box Office)
IBM (International Business Machines)
NATO (North Atlantic Treaty Organization)
BASIC (Beginner's All-purpose Symbolic Instruction Code)

●**EXERCISE 2** Correct the use of capitals in the following sentences.

1. Two of the computer programmers were reading *New Uses For An Old Printer.*

17.8

2. Larry's new boss told him ALWAYS to sign the log when he left the building.
3. In Matthew's senior year at the university he enrolled in courses in Computer Science, Business, and History.
4. Three of the division heads were studying french to prepare for their six-week stay abroad.
5. Alexandria stated, "no one will finish his or her project after September 15th."
6. When the survivors appeared, their families thanked god for their safe return.
7. Franklin takes his vitamin a and b_{12} tablets every morning.
8. He is an iowan.
9. The x-ray showed a slight fracture.
10. Our export department hired her because she could speak german, spanish, and italian.

17.8 AVOID UNNECESSARY CAPITALIZATION.

Many words are capitalized that do not normally fall under the heading of proper nouns. Capitalize them when you use them in a title and do not capitalize them when you use them in general terms as common nouns.

Capitalized	Not Capitalized
Nancy enrolled in French, Algebra 201, and Fundamentals of Spoken Communication.	Nancy enrolled in a foreign language, algebra, and speech.
Her patron saint is Saint Cecilia.	She has a patron saint.

Note: More examples are found in Section 17.1.

●18

ITALICS

- Italics are the print version of underlining.
- Underline titles of long documents.
- Do not overuse italics.

Italics for typed and handwritten documents are indicated by underlining. Printed material has italic type.

Typewritten	**Printed**
The report was entitled <u>Five Ways to Cure a Computer Hangover.</u>	The report was entitled *Five Ways to Cure a Computer Hangover.*

Italics (underlining) are used in a number of writing situations.

18.1 ITALICIZE (UNDERLINE) TITLES OF SEPARATE PUBLICATIONS, NOT THOSE FROM COMPILATIONS.

Usually these are publications of considerable length.

a. Italicize (underline) titles of books.

EXAMPLES

> *Matter, Earth, and Sky*
> *A Brief History of Time*

18.1

b. Italicize (underline) titles of periodicals.

Periodicals refer to all magazines, journals, and newspapers.

EXAMPLES

> *Newsweek*
> *The Chicago Tribune*
> *The Chronicle of Higher Education*

c. Italicize (underline) titles of long musical works and long poems that have been published separately.

EXAMPLES

Musical Works	**Long Poems**
The Grand Canyon Suite	*Beowulf*
Pictures at an Exhibition	*The Wasteland*
Finlandia	*The Odyssey*

d. Italicize (underline) dramatized works: theatrical, televised, and cinematic versions.

EXAMPLES

> *Long Day's Journey into Night*
> *Masterpiece Theater*
> *Casablanca*

e. Italicize (underline) works of art.

EXAMPLES

> *Mona Lisa*
> *View of Toledo*
> *Nude Descending a Staircase*

f. Italicize (underline) titles of formal reports.

EXAMPLE

The Feasibility of Installing a Waste Water Treatment Plant in Beetown, Wisconsin

18.4

18.2 ITALICIZE (UNDERLINE) FOREIGN OR NON-ENGLISH WORDS AND PHRASES.

EXAMPLES

la belle monde	*coup d'etat*
s'il vous plaît	*aloha oe*

Note: Do not underline words that have become part of our everyday vocabulary. For example, foreign foods such as pizza, lasagna, tacos, and the like are not underlined. Consult a dictionary to be sure a term is accepted as part of the standard English vocabulary. Although Latin abbreviations are still underlined by some, the tendency is not to italicize them.

EXAMPLES

ibid., e.g., i.e., vs., f., ff., op. cit., viz.

18.3 ITALICIZE (UNDERLINE) SPECIFIC NAMES OF SHIPS, PLANES, AND SPACECRAFT.

EXAMPLES

Titanic	*Apollo I*
Air Force I	*Voyager II*

Note: Do not italicize names of trains and general classes of items.

EXAMPLES

Super Chief	DC-7
aircraft carrier	Boeing 707

18.4 ITALICIZE (UNDERLINE) WORDS, LETTERS, OR NUMBERS THAT ARE BEING EXPLAINED OR ARE READ AS WORDS, LETTERS, OR NUMBERS.

EXAMPLES

Words

The term *interface* has more than one meaning.

Letters

When she discovered she had used a *z* instead of an *x*, she was embarrassed.

Numbers

Pat's *7* looked like a *1*.

18.5   DO NOT OVERUSE ITALICS (UNDERLINING) FOR EMPHASIS.

Use italics only when stressing a word or an idea. Too much underlining or too many italics cause the reader to lose sight of the important elements in a document.

Weak	**Improved**
If we *really* want the contract, we *must* act *now*.	If we really want the contract, we must act *now*.

18.6   DO NOT ITALICIZE (UNDERLINE) TITLES OF MAJOR HOLY BOOKS OR LEGISLATIVE DOCUMENTS.

EXAMPLES

the Talmud	the Mayflower Compact
the Koran	the 17th Amendment
the Bible	the Constitution

●**EXERCISE 1**   Underline all words in the following sentences that need to be italicized.

1. Martin read The Double Helix during his lunch breaks.
2. Greenblatt loves to listen to Rhapsody in Blue.
3. Both Newsweek and Time ran the same cover story.
4. When Margarita first used a computer keyboard she had trouble distinguishing between a zero and a capital O.
5. During her leave of absence, Meredith toured the Queen Mary.
6. Both Sam and Tina forgot to define the term force in their reports to the stockholders.
7. Luther's favorite TV show was Hill Street Blues.
8. Albert misused ibid. in his footnotes.
9. Augusta subscribes to the Christian Science Monitor.
10. The cast of Ah, Wilderness! toured the East for three months.

●19

HEADINGS AND SUBHEADINGS

KEY POINTS

- Break up long reports with headings for ease in reading.
- Choose a clear format for headings.
- Use the format consistently throughout a document.
- Employ an outline format to simplify a long text.
- Use the same alpha-numeric sequence in a table-of-contents outline as will appear in the text.

Headings and subheadings are needed in many reports. Not only do they break up the monotony of paragraphs, but they serve to help the reader find a section of the report. Practices vary as to exactly how each heading and subheading is written. This section gives one system that is used frequently. Others that are used vary in capitalization, underlining, spacing, and centering.

When headings are placed in a report, they often appear as follows:

<u>WATER DISTRIBUTION STUDY FOR HINSDALE, ILLINOIS</u>
[report title]

WATER SUPPLY AND DISTRIBUTION SYSTEM
[major area heading]

EXISTING FACILITIES [major topic heading]

<u>Radioactivity and Treatment Techniques</u> [subtopic heading]

1. Radioactivity [sub-subtopic heading]
2. Treatment Techniques [sub-subtopic heading]

The rules for this system of headings are given below.

19.1 CAPITALIZE AND UNDERLINE TITLES OF REPORTS. CENTER THEM ON THE TITLE PAGE.

EXAMPLE

<u>WATER DISTRIBUTION STUDY FOR HINSDALE, ILLINOIS</u>

Note: Your title should be a descriptive one that allows your reader to see at a glance what your report is about.

Weak	Improved
<u>WATER STUDY FOR HINSDALE</u>	<u>WATER DISTRIBUTION STUDY FOR HINSDALE, ILLINOIS</u>
<u>OXYGEN STUDY</u>	<u>AQUEOUS SOLUBILITY OF OXYGEN</u>

19.2 CAPITALIZE MAJOR AREA HEADINGS AND CENTER THEM ON THE PAGE.

Major headings serve as the main headings in your "Table of Contents." On the pages where they appear in the text, place them three spaces below any preceding text and two spaces above the following text.

EXAMPLE

WATER SUPPLY AND DISTRIBUTION SYSTEM

19.3 CAPITALIZE MAJOR TOPIC HEADINGS AND PLACE THEM AT THE LEFT MARGIN.

Major topic headings serve as the subheadings in your Table of Contents. On the pages where they appear in the text, place them two spaces below and above any text.

19.6

EXAMPLE

EXISTING FACILITIES

19.4 UNDERLINE OR ITALICIZE SUBTOPIC HEADINGS, CAPITALIZE THE MAJOR WORDS, AND PLACE THE HEADING AT THE LEFT MARGIN.

Subtopic headings do not appear in your Table of Contents. They serve to further divide a long section. On the pages where they appear in the text, place them two spaces below and above any text.

EXAMPLE

Radioactivity and Treatment Techniques

19.5 NUMBER SUB-SUBTOPIC HEADINGS, CAPITALIZE THE MAJOR WORDS, AND PLACE THE HEADING AT THE LEFT MARGIN.

Sub-subtopic headings serve to even further divide a very long section. Place them two spaces below and above any text. Like subtopic headings, they do not appear in your Table of Contents.

EXAMPLE

1. Radioactivity
2. Treatment Techniques

19.6 IF USING AN OUTLINE FORM, BE SURE TO FOLLOW THE CORRECT RULES FOR OUTLINING.

a. Capitalize, underline, and center titles as they are in Section 19.1.

b. Use Roman numerals for all major headings, capital letters alphabetically ordered for all major topic headings, Arabic numerals for all subtopic headings, and lowercase letters alphabetically ordered for all sub-subtopic headings.

19.6

EXAMPLE

I.
  A.
    1.
    2.
        a.
        b.
        c.
  B.
II.

Note: Some companies prefer an Arabic numbered outline. This form is used by the military in aircraft specification manuals.

EXAMPLE

1.
  1.1
  1.2
    1.2.1
    1.2.2
      1.2.2.1
      1.2.2.2
          a.
          b.
  1.3
2.

c. If a subject needs only one division, do not outline it.

You must have two entries to include them in an outline. Thus, if you have a *I*, follow it with a *II*. The same rule holds true for letters of the alphabet and for Arabic numerals.

● 20

SUPERSCRIPTING AND SUBSCRIPTING

KEY POINTS

- Distinguish between superscripting and subscripting.
- Use superscripting in documentation and equations.
- Use subscripting mainly in equations.

Superscripting refers to inserting numbers, letters, or symbols one-half space *above* the printed line. Subscripting refers to placing numbers, letters, or symbols one-half space *below* the printed line. Superscripting is used both in footnoting and in equations. Subscripting is most commonly found in equations.

20.1 IN DOCUMENTING WRITTEN DATA, USE SUPERSCRIPTED NUMBERS.

Placing the number of the footnote or endnote one-half space *above* the printed line makes the number easy to locate.

EXAMPLES

A comparison of the after-tax profits with and without the financing for pollution-control equipment makes it possible to quantify and analyze such an effect.[1]

Gumbel recommends use of log extreme-value-probability paper.[2]

---

[1] Environmental Protection Agency, *Choosing the Optimum Financial Strategy: Upgrading Meat Packing Facilities to Reduce Pollution*, EPA Publication No. 625/3—74—003 (Washington, DC: US Government Printing Office, 1973), p. 13.

[2] E. J. Gumbel, "Statistical Theory of Droughts," *Proceedings: American Society of Civil Engineers* (SSp. No. 439), May 1954, 60, pp. 1—19.

20.2 IN DOCUMENTING GRAPHICS, USE SUPERSCRIPTED LETTERS OR SYMBOLS.

Graphics are documented separately from the rest of the text in which they appear; therefore, they employ letters or symbols instead of numbers. The most common symbols used are those that appear at the top of the numbers on the typewriter—for example, #, &, *, and %.

EXAMPLE

TABLE 1. NITROGEN UPTAKE RATES FOR SELECTED FORAGE CROPS[a]	
NITROGEN UPTAKE RATE	
Crop	*kg/ha-yr*[b]
Coastal bermuda grass	400
Reed canary grass	340
Alfalfa	300
Quack grass	240
Rye grass	200
Kentucky bluegrass	200
Sweet clover	180
Tall fescue	160
Brome grass	130

[a]Values represent lower end of ranges presented in Table 12 and are intended for use in Equation 2.
[b]kg/ha-d = 0.893 lb/acre-d
Source: EPA Process Design Manual, Land Treatment of Municipal Wastewater.

20.3 IN WRITING EQUATIONS, BE CAREFUL TO USE SUPERSCRIPTING AND SUBSCRIPTING CORRECTLY.

Equations have to be accurate. It is important that all superscripting and subscripting be double-checked.

20.3

EXAMPLES

$$a^2 + b^2 = c^2$$

$$l_e(1 - e_e) = L_1(1 - e_1)$$

$$(Q_1 + Q_2)T_m = Q_1T_1 + Q_2T_2$$

DICTION

●21

TONE

- Use the formal tone in impersonal documents.
- Use plain, understandable language.
- Avoid jargon unless all your readers are familiar with the terms.
- Decide what tone to use, depending whether the readers are nontechnical, semitechnical, or highly technical.
- If your readers are at more than one technical level, adjust the tone to the least technical reader.

Tone is a term used in music—a sound of distinct pitch, quality, and duration. In your writing, the choice of words, phrases, sentence structure, organization, and development indicate your attitude toward the subject and toward the reader. This combined attitude is generally the distinct tone of the work.

In technical writing, your attitude toward the reader is usually classified as either formal or informal in tone. Both formal and informal tones follow all the grammatical conventions of written language. Formal tone, however, establishes an impersonal attitude toward the reader.

These two tones can be further distinguished as follows:

Formal Tone	Informal Tone
formal grammatical conventions	formal grammatical conventions
third person	first, second, or third person

impersonal	friendly
no contractions	contractions
longer, more complex sentences	shorter, simpler sentences

Because the informal tone is friendlier, it is used for most in-house communications, many business letters, and instructional writing. The formal tone is used for all formal proposals and reports, some business letters, most scientific publications, and various other technical documents such as abstracts and summaries.

The tone is also determined by the writer's attitude toward the subject. The document's content will be on one of three levels of technicality: nontechnical, semitechnical, or highly technical. The writer presents the technical-writing level appropriate to the technical level of the audience. The tone—choice of words, phrases, sentence structure, organization, and development—demonstrates whether the difficulty of the material and the comprehension level of the audience are aligned.

Often, the technical writer must present semitechnical and highly technical material to a nontechnical audience. In that case, the tone of the writing has to be more deliberately chosen so that the presentation of the material accommodates the audience's technical level.

21.1 THE FORMAL AND INFORMAL TONE USE DIFFERENT PERSONAL PRONOUNS.

a. When using a formal tone, use the third person.

Formal writing is generally impersonal. As a result, do not address the audience directly.

EXAMPLES

The tests confirmed that the computerized equipment on the SCV-101 needed to be reexamined before the aircraft could be flight tested. (*Not* "Our tests confirmed . . . ").

When adjusting the components, it is necessary to dismantle the shaft case. (*Not* " . . . be sure to dismantle the shaft case.")

b. When using an informal tone, use the second person, the implied *you* ("understood" second person), and the first person when appropriate.

In the more personal informal tone, when directly addressing the reader, use *you* and *I* (or *we*). When giving commands, such as in instructions, use the implied *you*.

EXAMPLES

second person: Our staff will meet with you and your management team sometime before next month.

implied *you:*   Move the belt to the left. ["understood" second person]

first person singular: I received 14 samples on May 21.
first person plural: We arrived at the airport at 1400 hours.

21.2 WHEN USING THE FORMAL OR INFORMAL TONE, FOLLOW CONVENTIONAL GRAMMAR RULES. FORMAL TONE, HOWEVER, HAS A MORE IMPERSONAL ATTITUDE TOWARD THE READER.

While the formal tone relies on the conventions of the written word, the informal tone allows the spoken word to influence the written word.

Formal Usage	Informal Usage
It is I who am responsible.	I'm responsible.
To whom should Tom send this letter?	Who should receive Tom's letter?
A person should use facts correctly.	Use correct facts.

21.3 KEEP THE LANGUAGE SIMPLE.

a. Use plain vocabulary.

With both the formal and informal tones, keep the language simple. Neither the formal nor the informal tone employs words that could interfere with understanding. On the left side of the following table is a list of words that may not be in everyone's vocabulary. Their simplified equivalents are listed on the right.

Pompous	Simplified
activate	start
aggregate	whole
amorphous	shapeless
antithesis	opposite

ascertain	find out
autonomous	independent
commencement	beginning
contiguous	near, touching
discourse	talk
effectuate	cause
elucidate	clarify
formulate	make
impairment	injury,harm
incombustible	fireproof
initiate	begin
optimum	best
reservation	doubt
requisite	needed
utilization	use
veracious	true
viable	workable
vitreous	glassy

b. Avoid jargon, which interferes with communication, unless the readers are very familiar with it.

In general, do not use jargon, the specialized language spoken among members of a particular group, in place of a clear expression of ideas easily understood by the average nontechnical reader. Jargon is only appropriate when all the readers of a technical communication are thoroughly acquainted with the specialized words. Examples of jargon are listed below.

Jargon	Improved
analytical model	formula
audio-playback output unit	loudspeaker
effective communications	clear writing
financial resource	money
human-resource training	training
integer	number
parameter	factor, variable

EXAMPLES

This is a full-featured asynchronous communications systems library providing interrupt-driven support for the COM ports; I/O buffers up to 64K; baud rates up to 9600; modem control and file transfer.

21.4

A related subject is metempsychosis, mentioned first in the cantos and expounded later in the lyrics. The personae device creates a series of forms that project themselves into the translucent sphere of the ego.

21.4 ADJUST THE TONE TO THE TECHNICAL LEVEL OF THE READER.

In technical writing, tone is determined largely by readers' technical knowledge of the subject. Readers may be classified as nontechnical, semitechnical, or highly technical in relation to the subject.

Nontechnical readers are often viewed as the general public, as well as many supervisors and managers; semitechnical readers as industrial technicians; and highly technical readers as engineers and scientists. These classes are general and need to be adjusted to individual situations. Someone can be highly technical in one field and nontechnical in another.

a. Use clear explanations, avoiding technical terms and jargon, when writing for the nontechnical reader.

The nontechnical reader has little or no specialized training in the field for which technical data are being presented. In communicating with a nontechnical reader, do not use jargon or a technical vocabulary—the words and phrases specialists ordinarily use to describe a concept, process, or thing—nor long, complex sentences. Give the nontechnical reader the background and explanations needed in order to understand the subject. Do not frustrate the reader with tedious explanations and abstract theories. Vary the tone with the circumstances, although normally the tone is serious, objective, and factual.

EXAMPLES

The silver-grey fleck of silicon, nicknamed "the chip," is, at its simplest, an electronic circuit. The fleck has tiny wires and switches etched, or patterned, onto it.

Diphtheria is caused by a special germ that usually enters the body through contact with someone who has had the disease. It results in the throat developing a thick membrane or skin that interferes with breathing.

b. Carefully select technical vocabulary and background material when writing for the semitechnical reader.

The semitechnical reader has some specialized training in the field for which technical data are being presented. Use some technical vocabulary and somewhat longer, more complex sentences. The semitechnical reader expects to be given the background information necessary to understand the subject, definitions, explanations, and interpretations. Employ an objective tone that is usually serious and informal.

EXAMPLES

Fusion, the kind of nuclear reaction that converts mass into energy inside the sun and other stars, is widely regarded as one of the most promising means of generating electric power for the next century and beyond.

Relative files—one of the two types of random-access files—allow you to find exactly the piece of data wanted from the files. They are more convenient for data handling because they allow you to structure the files into records and then the records into fields within those records.

c. When writing for a highly technical reader, use an understandable technical vocabulary and jargon the highly technical reader knows.

The highly technical reader has considerable specialized training in the field for which technical data are being presented. Use the jargon and technical vocabulary as well as the abbreviations, symbols, and formulas/equations of the particular field. The highly technical reader will understand the technical data with little additional background on the subject, or definitions, explanations, or interpretations.

Note: When addressing a peer group, vary the tone. Use informal wording, humor if appropriate, and criticism, depending on the situation.

EXAMPLES

The PMI OP-15 consists of precision BIFET operational amplifiers with bias current in the picoampere range; leakage current brings the input bias current up to the nanoampere level.

21.6

As a piece of textural intensity in both the poetry and prose of certain late-nineteenth-century writers, diphthongal /ay/ is aesthetic evidence of a constant texture-structure relationship between phonemics and complex metaphysics.

●**EXERCISE 1** In the following sentences, identify words and phrases that would be considered inappropriately technical or scientific by the nontechnical reader, then identify nontechnical words that would not be in the average nontechnical reader's vocabulary.

1. The cessation of demonstrably good marketizing methods has had the effect of placing this company subsequent to its competitor.
2. On the occasion of your employment determination, I make the suggestion that you attempt to definitize your prospects.
3. Modularity has been deemed necessary to optimize computer marketability.
4. It has been demonstrated by management that our current status is reported to be unacceptable to those in ownership of the company.

21.5 IF THE DOCUMENT WILL BE READ BY READERS OF DIFFERING TECHNICAL EXPERTISE, ADJUST THE TONE TO THE LEAST TECHNICAL READER OR PROVIDE DIFFERENT VERSIONS OF THE CONTENT.

Many technical writings have more than one type of reader. That is, they have a primary reader or readers to whom the writing is immediately directed, and they have secondary readers who have a tangential interest in the topics. Secondary readers may be those who are affected by the writing, who have to carry out its directives, who act as consultants for the primary readers, and so forth. In such cases, govern the tone by the reader with the least technical knowledge of the subject. With some documents, such as the formal report, approach the subject from several different technical levels by using methods such as the executive summary (see Section 34).

21.6 BE SURE TO KNOW THE READER'S PRIMARY NEED FOR THE TECHNICAL DOCUMENT.

Although the informal tone is most often used, formal documents require a formal tone.

a. Employ the formal tone for formal reports and other formal documents, impersonal business correspondence, most published scientific writing, and in communications to superiors or dignitaries.

In the formal tone, pay strict attention to formal grammar conventions, use the third person (*they, he, she*), and employ more complex sentences than in other types of writing.

EXAMPLES

The parties for whom the research was done should prepare to purchase more steel containers in order to accommodate the increased production.

Prior to the meeting, the president intends to greet the guests who will have convened by the pavilion.

b. As a general rule, use the informal tone in most other technical communications.

Use the informal tone in memos, general and personal letters, some scientific writing, in communications to fellow workers, and in instructional and manual writing. Address readers in the first and second persons, as well as the third person, rely on relatively simple sentences, and use grammar conventions, but allow the spoken word to influence grammatical decisions.

Weak	Improved
The aforementioned desires a priority position on the manager's appointment schedule. (too formal)	Bill Grey wishes to speak with the manager as soon as possible.
He is nothing but an aggravation to the guys who work up front. (too informal)	He irritates the employees in the retail office.

● **EXERCISE 2** Identify the following sentences as formal or informal in tone.

1. The continual additions made to the common stock of knowledge frequently effect such a complete revolution in the organization that the established patterns may be swept away in a single day.

21.6

2. I can promise you that a good suggestion will be immediately employed.
3. If you come up with an idea, I'll see if I can get the mining company to accept it.
4. This company is a limited arena for the demonstration of an employee's power.
5. Apprehension of the pattern of preferment in this company is incomprehensible.

●22

WORDINESS

KEY POINTS

- Remove weak sentence beginnings.
- Edit ruthlessly to remove all unnecessary words.
- Beware of wordy phrases.
- Turn clauses into phrases whenever possible.
- Review the revision to be sure it still says what you mean.

One of the most important characteristics of good technical writing is conciseness. You can attain conciseness by choosing words that are precise and by eliminating unnecessary words.

Precise prose requires factual and specific word choices that make the exact point instead of wordy approximations. General words have their uses, such as establishing categories of topics. However, writers tend to use too many general words, and most importantly, use them when specific ones are needed.

Weak	Improved
By copy of a change acknowledgement, we are requesting that our banking company effect the correction on the corporate records. To ensure proper follow-up of future notification regarding your investment, please send multiple requests to each company.	Please answer the questions on the enclosed change-of-address card so we can tell Bank of America where to send your dividend checks. You should also notify ADQ and Redad of your changed address.

22.2

22.1 USE A STYLE THAT IS PREDOMINANTLY FACTUAL AND CONCRETE.

Usually general and abstract words merely create the need to say the same thing again, only with specific words, if communication is to take place. General and abstract words do help you make summary statements, such as those for topics or a thesis. Too often, however, technical writers use general and abstract words where factual and concrete words are needed in paragraph development. For example, instead of saying that the board of directors approved a budget measure, you should say that the board of directors approved a 5% increase in managerial salaries for next year.

a. Use specific words.

Establish the subject precisely and immediately by using specific rather than general words.

General	Specific
electronic equipment	portable computer
measuring instrument	yardstick
kitchen utensil	spatula
office supplies	desk stapler
wrench	crescent wrench
drawing instrument	engineering compass

b. Avoid abstract or empty words that add little to a technical document.

Technical writing should be a precise form of writing. Do not use words that make the writing vague.

EXAMPLES

factor, aspect, facet, element, situation, in the nature of, consideration, wonderful, nice, circumstances

22.2 ELIMINATE UNNECESSARY WORDS FROM SENTENCES.

a. Remove weak sentence beginnings.

Weak sentence openers do not add to a reader's understanding of a topic. Always revise to eliminate such openings.

EXAMPLES

(There are) Many computer programs (that) solve problems without great complexity.
This (is a) problem (that) is solvable.
(It seems to me that) The entire plant needs to be redesigned.
(It is my opinion that) The lack of customers reflects a lack of advertising.

b. Avoid unnecessary repetition.

The use of different words to say the same thing probably comes from a wish to be emphatic. But such writing creates redundancies.

EXAMPLES

This toy is a novelty (that is new) on the market.
The new construction site is (in) close (proximity) to a fuel storage area.
She gave the city council her (personal) opinion of what needed to be repaired.
His (fellow) classmates told him that the crystals turned into (sweet) sugar.
The circular saw (is a saw that) rotates a toothed disk at high speed.
She wrote an autobiography (of her own life).
Operating a lathe is easier after (repeated) practice sessions.

c. Edit wordy phrases.

Many often-used phrases are wordy. Make concise substitutions for these phrases.

Weak	Improved
at this point in time	now
the reason why is that	because
until such time as	until
in the event that	if
due to the fact that	because
bring all this to a conclusion	conclude
in a great many instances	often

d. Remove excess words by making clauses into phrases.

Many times modifying material is placed in its own subordinate clause when only an adjective or adjective phrase is necessary. Never use a subordinate clause when the same effect can be achieved with a word or a phrase.

Weak	Improved
Many of the customers were unable to find the items that were on sale.	Many customers could not find the sale items.
The sales representatives who attend the conference will not receive the tax forms.	The sales representatives attending the conference receive no tax forms.

e. Use subordinate structures to reduce wordiness.

Compound sentences are composed of two independent clauses linked together in one sentence by a coordinating conjunction (*and, but, or, yet, so, for, nor*) because the content of each clause is strongly related to the other and equally important. If the information in one of the clauses is of lesser importance, put it into a phrase or dependent clause.

Weak	Improved
The company that bought the warehouse from John Smith marketed it and they sold it to the first governor of the state.	The company that bought John Smith's warehouse sold it to the state's first governor.
Several store managers said that they wanted to meet and they wanted to discuss policy on the issue.	Several store managers requested a meeting to discuss policy on the issue.
The engineer told his ideas to the draftsman and the draftsman then discussed the plans with the technician.	After he heard the engineer's ideas, the draftsman discussed them with the technician.

f. Remove relative and connecting words that are not needed to understand the sentence.

Many words used to connect or relate parts of a sentence are unnecessary because the sentence is clear without them.

EXAMPLES

The supervisors who are retiring believe (that) the store will close.
I think (that) he knows (that) Tom is going (to be able) to retire.

g. Remove other unnecessary words throughout the sentence.

Weak	Improved
In the case of the industrial technology major who wishes to become an executive, he or she will be expected to have experience at all levels of the company.	The industrial technology major who wishes to become an executive will be expected to have experience at all company levels.
There have been many times that I have driven to work and have spent a half an hour trying to locate a good parking space for my car.	Many times I have spent a half hour looking for a parking space before work.
I am requesting corrective measures be taken to make an amendment to this law, an amendment that would include tougher sentencing for uncooperative suspects.	I am requesting an amendment that would include tougher sentencing for uncooperative suspects.

●**EXERCISE 1** Remove all unnecessary words in the following sentences.

1. In my opinion, there are several considerations or factors that we on the committee must or should decide so that a decision on the subject will be made.
2. When he was writing or penning a feasibility study of a projected shopping mall, an area offering a wide variety of shopping, he came across new zoning laws that cast doubt on the availability of the already selected building site.

3. In the time of year that we call spring, he drove his car to the automobile dealer who sold cars to replace the catalytic converter located in the engine of the car.
4. There were five engineers who spent the weekend at a cabin in the wilds at the national forest preserve near the Wyoming town by the name of Cody.

●23

USAGE

KEY POINTS

- Use the informal style for most technical writing.
- Avoid slang, jargon, and "pop" words.
- Avoid clichés and trite language.
- Select single-meaning words with the appropriate denotations and connotations.
- Use specific language.
- Avoid sexist language.

Standard English recognizes three levels of language usage: formal, informal, and casual/colloquial. You must choose the words, punctuation, and grammatical structure appropriate to the situation and audience (see Section 21).

The formal style requires traditional grammatical choices, an extended but unpretentious vocabulary, and carefully constructed, complex sentences.

The informal style uses the rules of correct speaking rather than the traditional rules of correct writing.

The casual/colloquial style resembles everyday spoken language and as such admits some grammatical structures not admissible in either of the other two styles.

23.3

23.1 USE THE FORMAL STYLE FOR FORMAL REPORTS, CERTAIN SCIENTIFIC PUBLICATIONS, AND IN COMMUNICATIONS WITH DIGNITARIES AND SUPERIORS.

EXAMPLES

This report establishes the need to create an interdepartmental program to combine existing training programs and field experiences with computing experience and understanding. Training personnel will receive materials that will help them understand the microcomputer as it relates to all areas of the corporation.

The relatively slow compression is referred to as adiabatic, reversible, or quasistatic; the relatively fast compression as nonadiabatic, irreversible, shock or possibly diabatic. While the adiabatic gas law is found in many elementary texts, less commonly known is the corresponding gas pressure relation.

23.2 USE THE INFORMAL LEVEL FOR MOST OTHER TECHNICAL-WRITING TASKS.

EXAMPLES

The project's basic objective is to train faculty and staff members from the ten state campuses how to use microcomputers and selected software for application in their own classrooms. If staff members do not fill some sessions, we will invite staff from other schools to fill the vacancies. Joan Mannheim says her department has developed new writing programs for all personnel. We want to study these programs, though now it doesn't seem that the new programs revamp the old ones enough to be considered for company adoption.

23.3 AVOID THE USE OF THE COLLOQUIAL STYLE EVEN WHEN THE READER IS A FRIEND OR RELATIVE.

Occasionally you will direct a technical communication toward someone you know well. In such instances, the informal style is appropriate.

Casual/Colloquial	Informal
Jim, you asked how come I hogged all the copies of the report. Answer—'cause I forgot.	Dear Jim, You asked me why I did not give you a copy of the report. The answer is that I forgot.

Jane, where are you at in getting that financial statement ready?	Dear Jane, How much more work and time will the financial statement require?

Note: Both casual and colloquial speech tend to be figurative and imprecise.

●**EXERCISE 1**   Identify each of the following sentences as formal, informal, or casual/colloquial style.

1. They discussed the newest information on the nutritional needs of salmon in fish hatcheries.
2. However one may conceive the effect of these propositions, one must consider that it is predicated on the assumed motion of a particle in a conic-section orbit.
3. The "heat is on" now for research in solar energy, while money for nuclear energy study has been laid by.
4. Money for research on solar energy is readily available, but so is money for the study of nuclear energy.
5. The school was an academic horror with its high-falutin attitude and miserable standards, but me and my friends put up with it.

23.4   AVOID THE USE OF SLANG IN TECHNICAL COMMUNICATION.

Slang is usually inappropriate for technical writing. Slang words and expressions usually employ familiar words in new ways, often figurative ways, but also can be coinages. Slang usually comes in and goes out of style quickly, although certain expressions such as *bus* and *mob* eventually may be accepted as standard English. Slang is colorful and often humorous, but it is also figurative, imprecise, and can be unknown to the reader. Avoid using it in technical writing.

Weak	Improved
He was wasted after studying all night.	He was exhausted after studying all night.
As soon as she knew the scoop, she blew town.	As soon as she knew the police had arrested Ralph, she disappeared.

Getting in the final report was a mega ordeal.	Getting in the final report was an ordeal.
The manager was knocked into outer space by the idea.	The manager thought the idea was excellent.

●**EXERCISE 2** The following sentences, basically informal in style, are written for the nontechnical reader. Identify any slang, jargon, or inappropriate technical terms in each.

1. I am not as upbeat as the professor is about eliminating trial-and-error innovation.
2. Unlike paged virtual memory, disk-swapping does not increase the number of applications that concurrently execute.
3. The high school dropout who is an underachiever can try collegiate matriculation for two trimesters to see if he can get with it.
4. In focusing on the water-cooled thermal-conduction module, a company may overlook the air-cooled packaging available in certain large computer series.

23.5 AVOID CLICHÉS AND TRITE LANGUAGE.

A cliché is a frozen or set expression, such as "tried and true" or "looking for a needle in a haystack," that people have heard so often that it looks "tired" and may fail to communicate any meaning at all. Trite language is a combination of words heard together so often that the phrasing has become stale, such as "It is with pleasure that I welcome you," or "you may or may not know," or "We have voiced our opinion." Trite language is usually less colorful than clichés that involve figures of speech, such as "mad as a wet hen," or rhetorical arrangements such as "for better or for worse."

Weak	Improved
May I remind you that . . .	Remember
All things come to those who wait.	Perseverance is an asset.
Better safe than sorry.	The investment does not have sufficient guarantee.
Nothing ventured, nothing gained.	Most profit and success are bought with risk.

I would like to thank you . . .	Thank you.
Please find enclosed . . .	Enclosed is a . . .

23.6 WRITE WITH PRECISION BY CHOOSING WORDS THAT SAY EXACTLY WHAT IS MEANT.

Writing with precision means knowing the denotation, or dictionary meaning, of a word, as well as its connotation, or positive or negative emotional meaning. Gain precision by choosing specific words instead of abstract and general ones that have vague, broad meanings. Use direct and economical diction.

a. Be sure of the denotation of all your word choices.

Dictionaries give the denotative meanings of words by identifying exactly what they represent. Technical writing must be accurate; choose words for their exact meanings.

Weak	Improved
The country suffered from a trade *unbalance*.	The country suffered from a trade *imbalance*.
One rumor and the company stock *pummelled* on Wall Street.	One rumor and the company stock *plummetted* on Wall Street.
The CEO's resignation speech was *heartrendering*.	The CEO's resignation speech was *heartrending*.
After his automobile design failed to meet industry standards, the engineer felt like a *piranha*.	After his automobile design failed to meet industry standards, the engineer felt like a *pariah*.
The bonus had no *affect* on him.	The bonus had no *effect* on him.
A good juror should be *uninterested*.	A good juror should be *disinterested*.

b. Select words with appropriate connotations.

The connotation of a word refers to certain emotions or associations that accompany it. Some words imply a negative, some a positive, meaning. Other words are simply associated with certain contexts more than others and, therefore, sound strange when used with markedly different vocabulary.

Weak	Improved
Certain legislations used to be followed in order to plane wood well.	Certain rules need to be followed to plane wood well.
Make a precise 1-inch deep, 1-inch long hack in the middle of the rod.	Make a precise 1-inch deep, 1-inch long cut in the middle of the rod.
From the airport, the governor wheeled the president to the capitol.	From the airport, the governor drove the president to the capitol.
Wanting to skip school with her friends, the girl told her mother a fiction.	Wanting to skip school with her friends, the girl told her mother a lie.

c. Avoid ambiguous words and phrases.

Sometimes a sentence has more than one meaning because of the placement of its phrases and clauses. A single word also can have more than one meaning, thus confusing the reader.

Weak	Improved
The students spaced themselves out for the examination.	The students took alternate seats for the examination.
People who talk about committing suicide need understanding and encouragement.	People who talk about committing suicide need understanding and kindness.
On Easter Sunday, the children each lay an egg on the altar.	On Easter Sunday, each child places an egg on the altar.

d. Use concrete words and phrases to enhance the factual content of a technical communication.

People of all languages have long understood that some words represent concrete things—words like *rock*, *tree*, and *water*—and others do not, like *wisdom*. The word *abstract* comes from a Latin word meaning "removed from," and was generally understood to refer to something removed from reality. Therefore, words like *tree*, *water*, *apple*, and *hair* were considered concrete, and words like *judgment*, *personality*, *happy*, and *brave* were considered abstract.

Semanticists, scholars who study language usage, have shown that levels of abstraction for seemingly concrete words exist as well.

They demonstrated that for a word to remain concrete, it must truly refer to a specific *thing* that can be pointed out. To them, a general word like *tree* means all the different types of trees and is therefore abstract, whereas a word like *oak* refers to a particular type of tree that can be pointed out. To be truly specific, a word must refer to or name a particular thing that can be pointed to. This new understanding adds a great deal to society's comprehension of what scientific knowledge is, since scientific language, by definition, is concrete and factual (see Section 22.1, p. 138).

Hence general words, though they may seem concrete, are abstract to varying degrees. The semanticists invented a graphic device called the "ladder of abstraction," which demonstrates how a series of related words becomes more and more abstract.

To demonstrate, we will start with a phrase that is as concrete as possible; that is, it refers to one thing that could be perceived with the senses if one needed to show another person exactly what the phrase describes. As we move *up* the ladder of abstraction, we remove the words describing the particular subject and change the remaining words into terms that categorize more and more items but that also continue to include the one subject.

EXAMPLES

hunting dog
retriever dog
golden retriever dog
Sammy Smith's golden retriever dog named Fido

writing instrument
hand-held writing instrument
pencil
automatic pencil
Pyramid automatic pencil
blue Pyramid automatic pencil
Joannie Johnson's blue Pyramid automatic pencil

The importance of this "ladder of abstraction" concept to technical writing cannot be overestimated. Too many technical communicators tend to believe that if a word is what has traditionally been called "concrete," then it is specific and factual. The ladders given here show how little specific information general phrases like *hunting dog* and *automatic pencil* provide and demonstrate how many

23.6

different items can be categorized by words at this level of abstraction.

For example, a person cannot really point out a *flower* by pointing to a *rose*, for then all flowers would seem to be roses. On the other hand, one can point to a rose, no matter what its size or color, and establish what the word means. Technical writers should attempt to use very specific language to communicate factually.

Note: General, abstract language is needed for technical definitions and descriptions that identify a new specific item as part of a general class of things with which nontechnical readers would be familiar. There are other good uses for general, abstract language, but as a rule the technical communicator should try to be as specific as possible.

Weak	Improved
Use a hammer to remove the nail.	Use the nail-puller end of a claw hammer to remove the nail.
The lenses help correct her poor vision.	The concave lenses help correct her nearsightedness.
With a brush, go over the paint.	With a size 6 camel-hair brush, make long continuous strokes from the top of the board to the bottom until the oil-base paint has no ridges in it.
The person in charge said to make out a form and wait.	The personnel manager said to answer questions 1–8, and then take a seat in the office until she returned.
The magazine told her little about the advances of science.	*Scientific Advance* magazine had only two articles a year on heart surgery techniques.
The car wouldn't run.	The Chevy Impala had a sticking butterfly valve in its carburetor.

●**EXERCISE 3** Rewrite the following sentences using specific terms (ones low on the "abstraction ladder") to replace general, abstract terms.

1. An experienced writer who has published many writings should talk with people who want to learn to write.
2. The machine began to make a noise, so we brought in some tools and began to repair it.
3. Turn the handle a little bit to the right so that it can still jiggle around some.
4. Letters sent to some of the clientele are intended to increase business.
5. Drill another hole next to the one nearest you and then put the nail in it.
6. I sent a communication to you to request that members of the organization work on establishing new policies.
7. Due to the weather, the plane was delayed leaving the airport.
8. They were the administrators during the last several crises.
9. Plans are being made to forward the funds to several new departments.
10. It will take a percussion tool to break up this old floor.
11. The office staff is always supplied with several stationery items out of this budget.
12. A man called asking if we would be interested in viewing some educational software he has.

23.7 AVOID ELEGANT VARIATION.

Elegant variation occurs when a writer needs to repeat a key noun in a paragraph and attempts to find substitutes for the word. However, repeating the word or using a pronoun for it instead gives the paragraph unity and coherence, making it easier for the reader to comprehend the material quickly.

Weak	Improved
The family enjoyed having dinner together. The homogenous group was home at last, and all these relatives had fun.	The family enjoyed having dinner together. They were home at last, having fun.
Oscar Hummell is the design engineer for the new plastic motor. According to Hummell, this plastic wonder won't wear out, either. The revolutionary device will be lightweight as well.	Oscar Hummell is the design engineer for the new plastic motor. According to Hummell, the new motor won't wear out. It will be lightweight as well.

23.7

The assistant manager is being transferred to Houston. The dynamic administrator has clearly shown outstanding capabilities and this up-and-coming woman will go far.	The assistant manager is being transferred to Houston. She is clearly outstanding, and everyone believes she will go far.

●**EXERCISE 4** Rewrite each sentence, removing all jargon, trite phrases and clichés, and elegant variations.

1. Last but not least, the committee should conjugate the possible actions and decide previous actions are water under the bridge.
2. Regarding the matter of burning the midnight oil, the Smith Associates say the meeting must end by 10 pm. This fascinating business group always keeps its head on straight.
3. The committee provided input data, and we provided the parallel circuitry needed to implement the plan in the foreseeable future.
4. At the present point in time, the cost effectiveness pay off evaluations can establish meaningful criteria for the corporate structure.
5. The company has, in a manner of speaking, accessed new parameters in redesign possibilities.
6. Due to circumstances beyond my control, I had to sell my Camaro '84 A-28, 305 V8, new rads.
7. While it is better to be safe than sorry, I tender my resignation.
8. Despite the fact that in the majority of instances this doesn't happen, I am very sorry to report that this time it has.
9. If we switch to the metric system, grandmothers will proclaim that a dyne of prevention is worth a newton of cure instead of an ounce of prevention is worth a pound of cure.
10. It has been a pleasure to be of assistance and we thank you for your cooperation.
11. A chromatic harmonica in the key of G offers about forty fairly standard frequencies in the range of 200 to 2000 Hz.

●**EXERCISE 5** Rewrite the following sentences, removing instances of elegant variation, and substituting pronouns for the initial noun.

1. The banana is a good source of the mineral potassium. The golden fruit should be added to everyone's daily diet. Literally a potassium bomb, it is a bonanza of health.
2. Having purchased a disk drive, Paul knew that these most valuable computer accessories should come in pairs. The wondrous rectangular boxes in duplicate tripled the computer's capabilities.
3. Everyone wanted to surprise the new president with the new computer system. However, the new system was too large to avoid attention. The system was noticed by the president soon after he entered the building.

23.8 DO NOT USE SEXIST LANGUAGE.

All discriminatory language should be eliminated from communication, but sexist language is the most difficult to eradicate because it is so pervasive. Eliminate sexist usage of nouns, titles, expressions, and pronouns in technical documents.

a. Avoid sexist nouns.

Weak	Improved
manpower	human resources, work force
common man, average man	common person, citizen, worker
mankind	humans, humanity, human race, people
modern man	modern age, modern society, modern civilization
white man	caucasian
working man	worker, wage-earner

b. Avoid sexist titles.

Weak	Improved
chairman	chairperson, chair, presiding officer
congressmen	members of Congress, representatives
fireman	firefighter

headmaster	principal
policeman	police officer
salesman	sales agent

c. Replace sexist expressions.

Weak	Improved
founding fathers	pioneers, founders
gentleman's agreement	informal agreement, oral contract

d. Avoid generic sexist pronouns.

Weak	Improved
Each nurse treats her patients the best she can.	Each nurse treats patients with the best care.
A careful secretary consults her dictionary often.	A careful secretary consults a dictionary often.
Everybody needs his own space.	All people need their own space.

● **EXERCISE 6**   Remove the sexism from the following sentences.

1. The designer's craftsmanship was well known.
2. The night watchman called the policeman over to the warehouse door.
3. Each of the nurses felt that she was underpaid.
4. The founding fathers entered into a gentleman's agreement before they signed the official charter for the national organization.
5. The inventor produced a more refined engine in the hopes of helping mankind.
6. The NASA administrators prepared to choose astronauts who would man the ship.
7. I wonder who will become the new chairman of the executive council.
8. When a doctor is chosen for the company, he should have extensive experience.
9. All of mankind waited for news of the war.
10. What girl will they hire for the new secretarial position?

EFFECTIVE SENTENCES

●24

DANGLING CONSTRUCTIONS

KEY POINTS
- Make sure modifiers relate clearly to only one part of the sentence.
- Be certain that dependent clauses and phrases have a word close to them in the sentence to modify.

Dangling constructions are individual words or groups of words that do not clearly modify any word in the sentence and thus are left without any support within the sentence. The dangling construction may seem to modify the wrong word or two different words at once—it is said to "dangle" uncertainly. Since technical writers strive for accuracy and clarity, dangling constructions, which can be single modifiers, phrases, or clauses, interfere with both objectives.

24.1 AVOID SQUINTING MODIFIERS.

Some modifiers may seem to modify two parts of the same sentence. Often these words or phrases are placed in the middle of a sentence and can modify what goes either before them or after them. Hence they are called *squinting* modifiers. To solve such a problem, move the troublesome modifier to clarify what is being modified.

Weak	Improved
She only formatted the disk.	She formatted the disk only.
The club was open to members only from Monday to Friday.	The club was open from Monday to Friday for members only.

They failed completely to close the streets.	They failed to close the streets completely.
I was told now the report is due.	I was told the report is due now.
The foreman promised every day to change the chemicals.	Every day, the foreman promised to change the chemicals.
Swimming too often exhausts me.	I become exhausted when I swim too often.

24.2 AVOID DANGLING PHRASES.

Sometimes modifying phrases do not clearly modify any word in the sentence. The most common type of dangling construction occurs when an introductory phrase does not modify the subject of the main part of the sentence, as in this example: "Having torn the exam in two, it could not be submitted to the instructor." Here the introductory phrase, "Having torn," says that an action was accomplished, but the sentence does not clearly identify the actor.

With this kind of sentence structure, the introductory phrase needs to attach itself to a subject that did the action, usually the subject of the main part of the sentence, as in this improved version: "Having torn the exam in two, the student could not submit it to the instructor."

Weak	Improved
After finishing the presentation, lunch was served.	After the engineer finished the presentation, the caterer served lunch.
Looking out the window, the factory was ugly.	Looking out the window, I saw the ugly factory.
My tent collapsed when playing in the yard.	My tent collapsed when I was playing in the yard.
When told the bad economic news, pandemonium broke out.	When told the bad economic news, they broke into pandemonium.
Having heard of the free tickets, plans were made.	Having heard of the free tickets, the seniors made plans to attend the circus.

24.3

24.3 REPAIR SENTENCES WITH DANGLING DEPENDENT CLAUSES.

Modifying clauses can become detached from the word they modify. In addition, some clauses, called elliptical because assumed words are omitted from them, fail to find a subject to modify, as in the following example: "When calling members about a meeting, they are hard to catch at home." This sentence can be corrected in two ways:

1. When I am calling members about a meeting, they are difficult to reach at home.
2. When calling members about a meeting, I find it difficult to reach them at home.

Weak	Improved
We brought bread in a small store that cost 65 cents.	In a small store we bought bread that cost 65 cents.
Mrs. Smith introduced John to Jane before she was divorced.	Before she was divorced, Mrs. Smith introduced John to Jane.
When covered with roses, I find the trellis beautiful.	When the trellis is covered with roses, I find it beautiful.
After hiding out for a week, only surrender saved his life.	After hiding out for a week, he had to surrender to save his life.

● **EXERCISE 1** Improve the following sentences that have unclear modifying words or structures.

1. After fixing his bike, the contest continued.
2. The secretary only typed the first page.
3. We began the new system totally unaware of mechanical problems.
4. Jane visited Harriet when she was in the hospital.
5. While inspecting the new pumping station, clothes were splattered with water.
6. When forced to abandon the house, there was no place else for him to go.

7. We ran six laps after the work day that exhausted us.
8. Even when trying to lie, the jury believed him.
9. While hidden in the long grass, the lawn mower almost ran over the mice.
10. Bill notified Jim he had been told to move to a subordinate position in the business.

●25

MOOD

KEY POINTS

- The indicative mood, which conveys statements, is the most common sentence form.
- The imperative mood always has an understood "you" subject.
- The subjunctive mood is used to indicate that something is contrary to fact or to indicate certain formal procedures.

Mood refers to how the verb functions in a sentence. The three moods are asserted in declarative statements or questions (indicative mood), in commands or requests (imperative mood), or in suppositions, hypotheses, recommendations, or conditions contrary to fact (subjunctive mood). Technical writers use the indicative and the imperative moods frequently and the subjunctive mood occasionally.

25.1 USE THE INDICATIVE MOOD IN SENTENCES THAT CONVEY INFORMATION OR ARE FACTUAL.

The most-often-used mood of verbs is the indicative. It is the verb form used in ordinary factual statements and questions.

EXAMPLES

indicative mood statement: The bearings are dry.
                            The training session finished early.

indicative mood question: When will the contract be signed?
indicative mood exclamation: That air shaft is hot!
indicative mood statement: If Bob was at the construction site, he
left a note there for me.

25.2 USE THE IMPERATIVE MOOD WHEN WRITING INSTRUCTIONS AND OPERATING CYCLES AND ON OTHER APPROPRIATE OCCASIONS.

Verbs in the imperative mood express a command, suggestion, request, or entreaty. The subject of the imperative verb is always the implied *you*, as in the following: "Pick up your books." The sentence means *you* pick up your books, but the subject *you* is not expressed. The imperative mood allows the technical writer to address the reader directly in order to communicate emphatically as well as to adopt the reader's point of view in describing a procedure.

EXAMPLES

imperative mood command: Place the lever on the left side of the pad.
Start the motor by moving the switch to the left.
imperative mood request or entreaty: Tell me if I am to sign the report also.
Lend me five dollars until Monday.
Please substitute our specifications for yours.

Note: The imperative mood is very important to technical writers because it promotes direct, economical sentences with an informal tone.

25.3 ALTHOUGH MOST USES OF THE SUBJUNCTIVE MOOD HAVE DISAPPEARED, USE IT WHEN APPROPRIATE.

The subjunctive mood is used for sentences that express something contrary to fact, whether stated conditionally, hypothetically, or purely imaginatively, and for certain formal statements of parliamentary procedure. The subjunctive mood form, once very important in grammar, is barely retained today. However, in the two cases mentioned, the subjunctive mood is still used by technical and other writers. Distinctive forms for the subjunctive mood occur only in the present and past tenses of the verb *be* and in the present tense of other verbs used with third-person-singular subjects.

EXAMPLES

subjunctive mood contrary to fact: If I were the owner, I would hire another secretary.

I wish I were the sales representative for Asia.

You act as if Johnny were a criminal.

The assistant manager insists that he be put in charge of personnel.

If I were to pay the fine, I could go free.

subjunctive mood in parliamentary procedures or formal demands: I move that the committee disband.

I move that Sally be chair-person.

I recommend that the motion be carried.

If the amendment were to carry, the law would be greatly changed.

Note: Writer's often replace the subjunctive mood with the indicative mood by rewording the sentence. This probably occurs because of the writer's discomfort with the rarity of the subjunctive.

Subjunctive	Indicative
If I were the owner . . .	If I owned the company . . .
If I were the driver, I wouldn't act like that.	If I drove the car, I would act sensibly.
They recommend that he be dismissed.	The recommendation is that he should be dismissed.

●**EXERCISE 1** Identify each of the following sentences as in the indicative, imperative, or subjunctive mood.

25.3

1. Take those shoes back.
2. This meeting is two hours over schedule!
3. Try to look as if you were the Queen of Sheba.
4. Be it resolved that no children are allowed.
5. The football game ended in defeat.
6. Apply the lacquer with even strokes.
7. Please read me the printout.
8. If I operated that store, I would sell ice cream.
9. Give the foreman the raise he deserves.
10. The paint is not yet dry.
11. Do you think the paint is dry?
12. Please send this form back to the office.
13. Turn the wheel one full revolution to the left.
14. It is essential that he be honest.
15. If you were the vice-president, what would you do?

●26

VOICE

KEY POINTS

- Use the active voice most of the time.
- Remember that the active voice promotes direct and forceful sentences.
- Remember that the active voice also promotes conciseness.
- Use the passive voice only when absolutely necessary: when the actor is unknown, protected, or not emphasized.

The *voice* of a verb tells the reader whether the subject performs an action (active voice) or receives an action (passive voice).

Active Voice	Passive Voice
The manager (subject who performs an action) fired Bill Smith.	Bill Smith (subject who receives an action) was fired by the manager.

Because the subject performs the action, the active voice is direct and, therefore, vigorous and powerful. Active voice always uses fewer words than the passive voice, making it more concise.

The passive voice either names the actor in a prepositional phrase or omits the actor. The verb itself can never be constructed with only one word.

Much of technical writing describes work; the reader needs to know about the work but usually does not care who did it. Because of this emphasis on the work and not the worker, technical writers in the past used an abundance of passive structures. As the limitations of passive voice became better understood, they began using the active voice. Today, technical writers use the active voice whenever the actor is known. The active voice gives a stronger emphasis to all parts of the sentence.

26.1 USE THE ACTIVE VOICE FOR MORE DIRECTNESS AND LESS WORDINESS.

Passive Voice	Active Voice
The dinner always is given by the company president.	The company president always gives the dinner.
Most apartments soon will be available for rent by families with children.	Soon families with children can rent most apartments.
Some of the work was not submitted, so the financial report could not be completed.	The credit department did not submit figures and delayed the financial report.

26.2 REMEMBER TO USE ACTIVE VOICE EVEN WHEN THE ACTOR (GRAMMATICAL SUBJECT) IS NOT A HUMAN ACTOR.

The strong, active verbs still emphasize the work sufficiently.

Passive Voice	Active Voice
Almost all keyboard problems can be resolved by special software packages.	Special software packages can resolve almost all keyboard problems.
The cylinder is closed at both ends by thin metal caps.	Thin metal caps close the cylinder at both ends.
The lithium molecules are vaporized by heated wires.	Heated wires vaporize the lithium molecules.

26.3 USE THE ACTIVE VOICE FOR INSTRUCTIONS.

The active voice will make the instructions clearer and stronger.

26.4

Passive Voice	Active Voice
The paint should be spread evenly over the boards.	Spread the paint evenly over the boards.
The jar must be opened carefully so that spills are avoided.	Open the jar carefully to avoid spills.

26.4 USE THE PASSIVE VOICE ONLY WHEN NECESSARY: WHEN THE GRAMMATICAL SUBJECT (ACTOR) IS UNKNOWN OR NEEDS TO REMAIN UNKNOWN, OR WHEN THE SENTENCE'S EMPHASIS IS ABSOLUTELY NOT ON THE ACTOR.

EXAMPLES

The sample was somehow contaminated.
The owners said, "The embezzled money was returned," and refused to name the culprit.
The galaxies were formed an indeterminably long time ago.
The hospital not the clinic was sued by Mr. Smith.

●**EXERCISE 1**   Change the passive voice to active in the following sentences.

1. The company was told by the environmentalists to install a filter on two chimneys.
2. Workers should always change stations if they are told to by efficiency experts.
3. He should have been more determined to do what he had been required to do.
4. If the cases have been sold, there is no need to consider storage.
5. The payroll office has been working on a new plan for employees who have been fired.
6. The scandal was revealed by the president's secretary.
7. Fortunately, the funds were contributed by the banker.
8. The manager was suspended by the commissioner.
9. Five different candy bars were recalled by the company.

●27

PARALLELISM

KEY POINTS

- When sentences contain coordinate or serial structures, the structures should be similar in form.
- Use words to signal parallel structure.

Parallelism refers to grammatically equal sentence elements that indicate certain content material; items such as lists, series, compound structures, and related individual sentences should receive equal emphasis. The coordinating connectives are *and, so, but, or, how, for, yet.* (*So* and *yet* are often used as adverbs rather than as coordinating conjunctions. For example: "He goes so fast and we have not yet finished.") Correlative conjunctions such as *either/or, neither/nor,* and *not only/but also*) also link some parallel structures. At other times, a word or phrase may be repeated to denote or emphasize parallelism. Used appropriately, parallelism adds clarity and style to technical writing.

27.1 TO EMPHASIZE THAT CERTAIN CONTENTS OF A SENTENCE OR PARAGRAPH HAVE EQUAL IMPORTANCE OR EQUIVALENT FUNCTIONS, USE PARALLEL STRUCTURES.

Parallel structures are not only stylistically pleasing, but also clarify content by emphasizing that what is said in like form is of equal importance to the subject.

27.1

a. Use parallel words and phrases to demonstrate equal ideas.

Sometimes the parallel elements are single words, such as adjectives, or phrases, such as prepositional or participial phrases.

Weak	Improved
She was lovely and walked with grace.	She was lovely and graceful.
We heard stomping in the hallway and something stomped on the porch.	We heard stomping in the hallway and on the porch.
The three items on the agenda were as follows: the company picnic, how to distribute department awards, and the parking petition.	The three items on the agenda were as follows: the company picnic, the department awards, and the parking petition.
Taking orders, helping customers, and attempts to track down transportation are the major tasks.	Taking orders, helping the customers, and tracking down transportation are the major tasks.

b. Use parallel clauses to demonstrate content equivalency.

Sometimes the parallel elements are clauses, either independent clauses in a compound sentence or dependent clauses within a simple sentence.

Weak	Improved
He asked what we did and the people who were there.	He asked what we did and whom we met.
We shall meet with purpose and we'll get our work done.	We shall meet with purpose and we shall accomplish that purpose.
Because we are friends and also in conjunction with your talent, I commission you to do the art work.	Because we are friends and because you have talent, I commission you to do the art work.
The plans will continue in operation if you can file the	The plans will continue in operation if you can file the

proposal and also we may be able to convince the boss to fund the project.

proposal and if you can convince the boss to fund the project.

c. Use parallel sentences to demonstrate content equivalency.

Sometimes independent sentences are parallel. In these cases, the sentences begin with the same grammatical format and continue in parallel form, thus stressing the equivalency of their content. Often, because of this parallelism, semicolons are used to connect the sentences.

Weak	Improved
Success breeds success. If you fail, you will continue failing.	Success breeds success. Failure breeds failure.
	or
	Success breeds success; failure breeds failure.
If you are brave and wrong, people will adore you. Standing up for what is right won't necessarily get you applause.	If you are brave and wrong, people will adore you. If you are brave and right, people will abhor you.
	or
	If you are brave and wrong, people will adore you; if you are brave and right, people will abhor you.
The job of the chair is to direct committee action. On the other hand, the vice-chair should see that needed actions occur.	The job of the chair is to direct committee action; the job of the vice-chair is to carry out these directions.
	or
	The job of the chair is to direct committee action. The job of the vice-chair is to carry out those directives.

27.2 WHEN USING CORRELATIVE CONJUNCTIONS, BE SURE TO PLACE THEM NEXT TO THE PARALLEL STRUCTURES.

Correlative conjunctions—*both . . . and; either . . . or; neither . . . nor; not only . . . but also; whether . . . or*—usually connect parallel structures. Both parts of the conjunction pair should be placed with the parallel structures so as to introduce the parallel content.

Weak	Improved
John either should be promoted or fired.	John should be either promoted or fired.
I was not only interviewed by the staff but by the administrators as well.	I was interviewed by not only the staff but also the administrators.
This neither is to our advantage or disadvantage.	This is neither to our advantage nor to our disadvantage.
Whether you decide to come can't bother me very much.	Whether you decide to come or to stay home doesn't matter to me.

Note: When using the correlatives *not only/but also*, don't forget the *also* and keep it next to *but.* Many people fail to recognize that these are paired correlatives and write: "He not only paid the fee but the fine." The sentence should read: "He paid not only the fee but also the fine." The second sentence is clearer and more markedly parallel.

27.3 REMEMBER TO REPEAT THE WORDS THAT MARK PARALLEL STRUCTURES.

Weak	Improved
It is better to give than receive.	It is better to give than to receive.
We will beat them in the air, land, and sea.	We will beat them in the air, on the land, and on the sea.
Most students drop out of school because they are making bad grades and parents don't encourage them to stay.	Most students drop out of school because they are making bad grades and because their parents don't encourage them to stay.

Early to bed and rise make a
person healthy, wealthy, and wise.

Early to bed and early to rise
make a person healthy, wealthy,
and wise.

●**EXERCISE 1**   Rewrite the following sentences so that they have
good parallel structures where appropriate.

1. Parents often don't realize how much college will cost and
   they have no idea of ways to pay for it.
2. I am convinced of their sincerity by their conduct, talk, and
   persistence.
3. Because we want the sales and also need them, we all
   worked overtime.
4. The manager was fair, honest, and hard work didn't
   bother him.
5. I can tell you that either Sharon does her work or she is
   fired.
6. Jim is not only dishonest but a poor student.
7. To sleep, to sleep; perhaps I'll have a dream.
8. Whether the tickets are expensive determines if we go.
9. He was charming, witty, and had wonderful manners.
10. The three necessary amendments involve the following
    improvements: committee structure, having two breaks a day,
    and when to turn in tax forms.

●28

SUBORDINATION

KEY POINTS

- Combine a series of short, simple sentences into a stronger sentence that indicates the proper relationship among ideas.
- Beware of sentences that contain three or more dependent clauses.
- Use subordinate clauses to write more concisely.

Most technical writers strive for shorter and less complex sentences than those in formal style. Even so, technical material is made clearer and more concise when the writer places less-important material in subordinate structures.

Subordination refers to the writer's need to show the relative importance of the ideas expressed within a single sentence. Clauses, which have subjects and predicates, contain more important information than phrases, which do not have subjects and predicates. There are two kinds of clauses: independent and dependent or subordinate clauses. Independent clauses are meant to carry the sentence's most important information, with the dependent (subordinate) clauses carrying the next most important information.

Subordinate clauses are modifying structures. They can serve as adverbs or adjectives. Adverbial subordinate clauses begin with subordinating conjunctions. The most common are *while, when, after, because, although, if, since, as, before, than,* and *unless.* Adjectival subordinate clauses usually begin with relative pronouns, which are *who, whom, whose, that, which, what, whoever, whomever, whichever, whatever.*

EXAMPLES

The Meadville commuter train, *which has run for fifteen years,* has been cancelled.

Before the dinner began at eight o'clock, the guests heard an address by the president.

The committee wrote the proposal and five people began work on the project *while John was on vacation.* (This sentence has two independent clauses, both of which tell equally important information.)

Because the formal report was long, the board rescheduled the hearing for Tuesday morning at nine, *before Sarah could protest.* (This sentence has two subordinate clauses, both of which contain secondarily important information.)

The manager, *who is from Los Angeles,* asked everyone in the department to contribute names from other communities.

Since the design was almost finished, the corporation decided to hire a new foreman *whose qualifications would facilitate production.*

Conciseness is achieved when independent sentences are combined by making use of subordination. Clarity is achieved when sentence clauses are placed with the correct amount of emphasis on ideas of unequal importance.

28.1 COMBINE SEVERAL SHORT, PROXIMATE SENTENCES INTO ONE SENTENCE USING INDEPENDENT AND SUBORDINATE CLAUSES.

The following are examples of weak sentences with all independent clauses and improved sentences using subordinate structures.

Weak	Improved
We measured the rod. Then we cut it in .25-inch pieces. We used a band saw to do the cutting. It is able to cut through very hard metal.	After we measured the rod, we cut it into .25-inch pieces with a band saw, which is able to cut very hard metal.
The swimmer entered the water at 6:00 am for the contest. She swam well for two hours. Then she began to tire. At the end of three hours, she was no longer in the lead.	After entering the water at 6:00 am, she swam well for two hours before she began to tire, but after three hours she lost her lead in the contest.

28.2

One should prepare as thoroughly as possible. Then the interview won't be half the ordeal one might expect.

In one prepares as thoroughly as possible, the interview won't be half the ordeal one might expect.

This is the new production engineer. She used to work for Bendix Corporation.

This is the new production engineer, who used to work for Bendix Corporation.

The formal report is now ready. It was written by Fred Wiesnyski. He wrote the previous report that Mr. Wallen liked so much.

Fred Wiesnyski, whose previous formal report Mr. Wallen liked so much, has now completed the new one.

28.2 BE SURE THE MOST IMPORTANT INFORMATION IS IN THE INDEPENDENT CLAUSES.

The main clause of any sentence is an independent clause and should carry the sentence's vital information. If the sentence is a compound sentence, then it has two (or maybe more) independent clauses that contain equally important information. Any subordinate clauses will contain information of secondary importance.

Weak	Improved
He did not feel well when he ruined the 400 pieces of aluminum.	He ruined 400 pieces of aluminum when he felt ill.
The report, which was beautifully illustrated with advertising ideas, was five pages long.	The report, which was five pages long, was beautifully illustrated with advertising ideas.
The light fixture, which was ruined, had been installed years ago.	The light fixture, which had been installed years ago, was ruined.
The mayor was holding a news conference when word was received that he had lost the primary.	While the mayor held a news conference, word was received that he had lost the primary.
The committee met in the old headquarters when they discussed bankruptcy.	When they met in the old headquarters, the committee discussed bankruptcy.

28.3 AVOID EXCESSIVE SUBORDINATION.

If a writer places one subordinate clause immediately after another, the sentence will not read smoothly and the subordinate clauses might be too far from the words they modify.

Weak	Improved
A company who wants better profits should not depend on the guidance of a chairperson who is on the board of several other companies who probably are taking time when that is impossible.	A company who wants better profits should not depend on aid from a chairperson with no time to spare.
The joke that Bob played on Tom was not funny because the teacher saw what Bob did and that caused Bob to be expelled.	The teacher asked the principle to expel Bob for the joke that he played on Tom.
Half of all the galaxies that are in the universe are members of a group or cluster that are of a similar size.	Half of all the galaxies in the universe are members of a group or cluster of similar size.
The plant that belongs to the Orion Corporation and that produces the most electricity and that serves our whole county is closing.	Owned by the Orion Corporation, the electric plant in our county that produces the most energy is closing.

● **EXERCISE 1**   Rewrite the following by increasing or decreasing the subordination necessary to produce a concise, clear sentence. In some instances you will have to decide what the sentence intended to say.

1. The farmers were upset by recent legislation. They felt the new laws hurt them even though they were said to help them. The farmers decided to seek a referendum to change the law.
2. The researchers who were gathered in the hall were trying to organize an all-company meeting that would take the place of the report that was the usual year-end task that was, however, only for upper-level engineers.

3. The building size was good. The land included a small river and access to an interstate highway. These were both necessary. The engineers were glad they had purchased the land.

4. Their plan, which consisted of flying the machine as a kite, had worked out as they had expected and as they had agreed was most beneficial to their company.

5. When the legislation protecting small business failed to pass, several senators supported it.

6. The major points of selling techniques were outlined. The training personnel made us act out specific selling situations. One trainee became angry with the person acting as a potential customer. He slapped her. The session came to an end.

7. After we discussed the time and after we decided on a date and although three of the ten people couldn't attend, we adjourned.

8. When she died of a heart attack after being in the water for four hours, the spectators were very shocked.

9. Hans felt that the supervisor had been too strict. He went to work prepared to confront him. Then Hans found out that his incompetence had caused a worker on the night shift to be injured.

10. While handball is a very strenuous sport, it is more demanding than ice skating.

● 29

UNITY

KEY POINTS

- Make sure each sentence is unified.
- Clearly indicate how the parts of a sentence are related to one another.
- Do not clutter a sentence with excessive details that hide its focus.
- Avoid mixed metaphors.
- Do not shift construction in the middle of the sentence.

Sentence unity refers to a sentence having one focus—clarity. In order for a sentence to be unified, all of its parts must be closely related and it must not have too many details or unclear constructions.

Unity problems occur not only because of what is said, but also because of what is not said. Sometimes a writer fails to show what logical connection exists between sentence parts. The writer knows what is *meant*, but the idea is incompletely presented to the reader.

29.1 DO NOT PUT UNRELATED IDEAS IN THE SAME SENTENCE.

Frequently, writers know the relationship between their expressed ideas but fail to inform the reader. Sometimes writers try to

condense their writing too much and put together insufficiently related information in one sentence. Other times, the relationship may
be written but not clearly.

Weak	Improved
Half the tools had not been replaced on the shelf, and the boss was on vacation.	Because the boss was on vacation, much work was left unfinished, such as the replacement of tools after their use.
John worked very hard, and his sister was to have an eye operation.	Because John worked very hard and saved his money, he could finance his sister's eye operation.
The technical-writing staff had sandwiches delivered, and they worked hard on the manual.	The technical-writing staff was working so hard on the manual that they did not take time out to go to lunch, so later they had sandwiches delivered.
The insurance company investigated the loss and told the owner about the new company rates.	After the insurance company investigated the company's second lost shipment in a month, they told the owner he was now in a new insurance bracket with higher rates.
After making 100 copies of the account ledger and asking to see the president, the secretary left the office.	After fulfilling the president's unreasonable request for 100 copies of the account ledger, the secretary quit.

29.2 DO NOT OBSCURE A SENTENCE'S UNITY WITH TOO MANY DETAILS.

It is false economy to load a sentence with too many details
because the reader will be unable to sort out the necessary information. Sometimes the writer should create more than one sentence; at
other times, the answer is simply to leave out unnecessary detail.

Weak	Improved
When Beth, who is a native of New York City and lived there for fifteen years, only recently becoming a resident of Wisconsin, became a safety engineer in the highway department, a body that had been employing safety engineers for 85 years, she decided she liked her new state.	Beth, a native New Yorker, decided she liked her new state, Wisconsin, when she became a safety engineer in the state's highway department.
After John, who had worked for the company for 15 years, had made a jig out of a piece of scrap aluminum weighing about a pound, he clamped it onto the milling table that Mildred had used and she began to line up the plastic pieces for drilling.	An old hand in the shop, John quickly made a scrap aluminum jig so Mildred could line up the plastic pieces on the mill for drilling.

29.3 AVOID MIXING METAPHORS IN A SENTENCE.

A metaphor is a figure of speech, which means, in general, that it is language used in an imaginative rather than a literal sense. In particular, a metaphor is a word that makes a comparison in a very direct way. Instead of the poet's saying that his love is *like* a red, red rose, he could have addressed his wonderful rose. The metaphor replaces words such as *like* and *as* by saying one thing *is* another.

Nouns are not the only part of speech that can be metaphors. Verbs can be metaphors when something does an action described in terms of something else. If we say that an angry man bristles, we have compared the man to an animal whose short, coarse, thick hair stands erect when the animal feels fright or anger.

Metaphors tend to become mixed in a sentence because more than one comparison is being made. A writer may fail to realize one of the comparisons is a metaphor especially in trite expressions like "making a mountain out of a molehill" or "horsing around." Then the sentence might contain two or more metaphors with conflicting images.

Weak	Improved
With his ear to the ground, he put one foot into the forest.	Listening carefully, he put one foot into the forest.
He jumped out of the frying pan into the fire when he skated on thin ice.	He skated on thin ice and turned the police's suspicion into certainty.
After monkeying around with his car, he decided to go to the bank and squirrel away a little cash.	Once his car was running again, he went to the bank to squirrel away a little cash.
The owner was the foundation of the company, but his secretary Tom Avery was the salt of the earth.	The owner, as the foundation of the company, had established many successful policies, but the practicality of his secretary, Tom Avery, made them work.

29.4 AVOID AWKWARD CONSTRUCTIONS THAT HURT A SENTENCE'S UNITY.

Many sentences set up a balance between the subject and the material in the predicate that defines or describes the subject, such as the following: "The foreman should be an experienced, skilled worker." The sentence should make an equitable statement immediately before and after the verb (in this case, *should be*) about what the subject is.

Weak	Improved
When the foreman asks workers to pay attention to him is a reasonable request.	The foreman's request that workers pay attention to him is reasonable.
The first step is the operator starts the machine. (This sounds as if the operator is the first step.)	The operator's first step is to start the machine.

Fiction is when the author invents part or all of a story.

Fiction is a literary work whose content, wholly or partially, is produced by the author's imagination.

29.5 AVOID MIXED CONSTRUCTIONS.

Awkward structures that destroy a sentence's unity and make the writer's thinking appear incoherent also can be viewed as shifts in construction (see Section 6). The writer of such sentences starts a sentence one way, but fails to complete the idea before moving in a new direction. Language study indicates that this is a frequent writing error.

Weak	Improved
After the retirement plan is set in motion works well for all employees.	The retirement plan, after it is established, will work for the benefit of all employees.
The two men were determined to learn to do computer programming was a year-long ordeal.	The two men were determined to learn computer programming, but did not realize it would be a year-long ordeal.

●**EXERCISE 1**   Improve the unity of the following sentences. You will have to decide what logical relationship the writer intended to express.

1. Sylvia was late to work and brought her sack lunch with her.
2. The poor dope couldn't see the forest for the trees, but felt there was a silver lining somewhere.
3. John called Sam to pick him up for work and the car wouldn't run.
4. Whenever Myrtle went in to a restaurant downtown is the time she wished she made more money.
5. Even though he always seemed to have his shoe on the wrong foot, he never tripped over his own tongue.
6. When the newspaper, which had been in business for more than a hundred years, hired the woman from New York who had been a college president, the quality of writing, such as

feature stories, news releases, advertisements, became much higher in that no mechanical errors were tolerated, unity was respected, and a higher level of vocabulary was expected.

7. Although Sarah was enrolled in an engineering school felt she had not received sufficient math.

8. The committee pleaded with the administration to let the ruling stand and also planned the annual dance.

9. The way to begin is when the workers strike, be ready.

10. The balance of the evening is to eat, drink, and be merry.

●30

ENUMERATION

KEY POINTS

- Use enumeration to emphasize important data.
- Use enumeration for formatting variation.
- Set off enumerated lists with some form of marking, such as numbers, asterisks, dashes, or bullets.
- Punctuate continuous enumerated lists as if the data were presented in paragraph form.
- Clearly incorporate informal tables into a text.
- Use noncontinuous enumerated lists for data that are boxed off from the rest of the test.
- Punctuate and capitalize noncontinuous enumerated lists for emphasis.

Technical writers often choose to display lists of figures, dates, personnel, important points, and the like by indenting the list and by numbering the items. Such enumerations are usually a continuation of the text and employ punctuation as if the data were presented in paragraph form. Sometimes they are not continuous with the text but give incidental or separate data. Then the enumerations have an introductory sentence and both the sentence and the list are inside a box.

Enumeration is a means of emphasizing material that might otherwise be buried in a paragraph. It also simplifies complex material by separating major points from the explanations and interpretations that remain in the paragraph. Finally, it contributes to an interesting format by breaking up the page.

30.1 NORMALLY NUMBER A LIST, ALTHOUGH ASTERISKS, HYPHENS, AND BULLETS ARE ALSO ACCEPTABLE.

EXAMPLES

Check the body of your letter for the following:

1. salutation		*salutation
2. closing	*or*	*closing
3. signature		*signature
4. typed name		*typed name

Check the body of your letter for the following:

-salutation		•salutation
-closing	*or*	•closing
-signature		•signature
-typed name		•typed name

30.2 PUNCTUATE ENUMERATED LISTS THAT ARE CONTINUOUS WITH A PARAGRAPH AS IF THE DATA WERE GIVEN IN PARAGRAPH FORM.

Frequently, technical communications incorporate some form of enumeration into a prose paragraph. In such instances the enumeration is always viewed as part of a paragraph and punctuated or capitalized accordingly. The introductory sentence, therefore, does not necessarily have punctuation immediately preceding the list. Items in the list are separated by commas as if they were written as a series in a regular sentence. A period ends the last entry in the list.

EXAMPLE

In order to keep the customer happy, I suggest that

1. all correspondence be answered promptly,
2. the customer be contacted by phone, and
3. a contract be forthcoming as soon as possible.

Let me know if I can do anything to help.

●**EXERCISE 1** Using the format for letters discussed in Section 31, compose a letter to a high school graduating class informing them of what to expect of college, or some aspect of college, and create a list as part of the letter. Make it continuous within the text.

30.3 VARY PUNCTUATION FOR ENUMERATED LISTS THAT ARE NOT CONTINUOUS WITH A PARAGRAPH.

Some enumerations are not given as part of a paragraph. Rather, they are separated from the rest of the communication and frequently have a box around them. They may precede or follow a text.

EXAMPLES

The recently-formed bid committee has the responsibility of investigating bid procedures. All meetings will be held in the Simpson Building at 8 pm. The date, place, and topic of next month's meetings are as follows:

DATE	PLACE	TOPIC
Feb. 8	Room 212	Smith County bid
Feb. 10	Room 214	Sawyer County bid
Feb. 14	Room 212	Mallar City bid
Feb. 20	Room 216	Baldwin Corp. bid

If a member of the committee must be absent, he or she must inform John Buehl, chairperson.

In case of a tornado alert, you should

1. Leave the room in an orderly fashion.
2. Walk quietly in a line to the nearest stairway.
3. Still walking quietly and in a line, proceed down the stairs to the basement.

If you follow these instructions, no one will be injured.

●**EXERCISE 2**   Using the format for letters discussed in Section 31, compose a letter to a group of your choice suggesting that the group undertake a particular activity. Explain your suggestion in the letter and give a list of several times, dates, and places, the activity could occur. Place the information in vertical columns, not continuous with a paragraph.

30.4 EMPLOY ENUMERATION AS PART OF SEVERAL VERTICAL COLUMNS IN AN INFORMAL TABLE.

EXAMPLES

The bids for the Sawyer County bridge have been opened and the results are as follows:

1. Waller and Keaton $1,708,900
2. Marquardt Co. $1,600,500
3. Brame and Sons $1,450,800
4. Dobble Inc. $,1,300,400.

As low bidder, Dobble Inc. was awarded the contract.

The commission's budget for this month includes the following items and allowed costs:

1) $ 25.00   for advertising,
2)   45.00   for legal fees,
3) 100.00   for paid assistants,
4)   15.00   for refreshments,
5)   30.00   for correspondence/copying, and
6)   30.00   for supplies.

The chairman makes a special request that the commission stay strictly within the allotted amounts.

●**EXERCISE 3**   Find a letter-to-the-editor in one of your local papers and rewrite it, incorporating an enumerated list of important information into a continuous informal-table form.

LARGER ELEMENTS

●31

BUSINESS LETTERS

KEY POINTS

- Use established formats for business letters.
- Write informally, avoiding pompous language.
- Address the reader ("you") as frequently as possible.
- Make letters as brief and relevant as possible.
- Address all business envelopes neatly and correctly.

Letters are one of the main written communications used in the business world. Most letters are relatively brief, relevant, informative, and, above all, clear. Learning to write effective letters should be a priority for anyone entering the world of business and technology. In general, a clear format, a pleasant tone, specific content, and simplicity characterize good letter-writing.

Letter writers may compose a report letter one day and a complaint letter the next, or they may write letters of adjustment, to the editor, or for transmittal. Whatever the type of letter, certain rules apply to all.

31.1 USE AN ACCEPTED FORMAT.

Type business letters on one side of unlined 8½-by-11-inch paper using a preestablished format. The two most popular formats used today are the *block* (see Figure 31-1, p. 193) and the *semi-block* (see Figure 31-2, p. 194).

Divide the business letter as follows: (1) return address or heading and date, (2) inside address, (3) greeting, (4) body, (5) complimentary close and signature, (6) notations, and (7) continuation pages.

The Return Address or Heading Begin with the return address, which includes your street address, city, and state, as well as the date you wrote the letter. If you are using letterhead stationery, type in only the date because the name of the company and its complete address appear in the letterhead.

If you are using the block form, type the return address flush with the left margin (see Figure 31-1). For the semi-block form, type the return address slightly to the right of center (see Figure 31-2).

The Inside Address Type the inside address two to four lines below the return address, flush with the left margin. Place the recipient's name and address in this part.

EXAMPLES

Mr. John Applegate   Miss Ann Lorenzo
Gateway Corporation   916 Allen Street
981 Marshall Road   Benton, WI 53811
Franklin, IL 61001

When you address the recipient as *Mr.*, *Mrs.*, *Miss*, or *Ms*, you run into two problems. First, you will encounter first names that can be either male or female. Common examples include Tracy, Kelley, Robin and Kim. In addition, you will find that abbreviations such as "A. H. Jorgenson" do not indicate sex. Second, you will not always know a woman's marital status because usually women are not identified by their marital status. Recently, *Ms* has been adopted by some companies; however, a number of women dislike this title. The best solution is not to use any title.

A number of businesses have dealt with this problem by dropping these titles and addressing people only by their name.

EXAMPLES

Jenny Luther	A. X. Sampson
711 Union Street	912 Sunrise Drive
Rockford, IL 61109	Madison, WI 53201

In addition to determining how to address the individual, you also will have to deal with business titles. Be sure to include the recipient's business titles in the inside address, even if an extra line is needed.

EXAMPLES

Jeremy Lightfoot, Manager	Rogene Armatto, Director
Swingway Corporation	Light Building Division
91123 Railroad Way	Mayfair Homes
Arthur, WI 53717	8888 Mayfair Road
	Waukesha, WI 53002

The Greeting or Salutation   Place your greeting or salutation flush with the left-hand margin two spaces below the inside address. Because of the problem with *Mrs.*, *Miss*, and *Ms*, as well as the problem with unisexual names, use one of two commonly-employed methods for handling the salutation.

1. *Address the person by his or her full name.*

EXAMPLES

Dear Polly Rector:
Dear William O. Justice:

2. *Leave the salutation out entirely.*

This second alternative solves another dilemma. When you address letters to organizations or to people whose names you do not know, your salutation becomes vague, such as "Dear Credit Manager," "Dear Sir or Madam," "Dear Personnel Manager," or "To Whom It May Concern." Simply eliminating these awkward and unnecessary salutations is a reasonable solution.

Another current change in salutations is the use of the comma. Some companies are replacing the colon with the comma to make the salutation less formal. The comma also is widely used in business letters written to acquaintances. Your company's policy will dictate which form to use.

The Body   Single-space paragraphs in the body of the letter and double-space between paragraphs. Do not indent the first line of a paragraph. You still will find some who do so in older formats, but it is unnecessary and is currently not considered good business-letter form. Section 31.2, p. 195, discusses the style in which the body should be written.

The Close or Complimentary Close   Type the close two spaces below the last line of the body of your letter, flush with the left margin in the block style (see Figure 31-1). In the semi-block style, type the close just to the right of center, lined up with the return address (see Figure 31-2). The two most popular closes are *Sincerely yours*, for a formal close, and *Sincerely*, for a less formal one. Other popular closes include *Yours truly* or *Very truly yours*, *Cordially* or *Cordially yours*, and *Respectfully* or *Respectfully yours*.

Signature   Leave four spaces for the handwritten signature. Follow this with your typed name. If you have a title, place it immediately below and aligned with your typed name.

EXAMPLE

> John Graboski
> President, Student Activity Board

Your handwritten signature should match your typed name and should always be signed in blue or black ink. In practice, people often sign only their first names when they know the person with whom they are corresponding.

Notations Only notations appear below the signature, flush with the left-hand margin in both the block and the semi-block forms. Several items may appear in these notations.

1. A notation indicating that you have enclosed something with the letter: *encl.*, *enc.*, or *enclosure(s)*.
2. Your initials and those of your typist's if someone else typed your letter. Your initials appear in capital letters, while those of your typist appear in lowercase letters. Examples are *FF/jk* or *FF:jk*.

Note: This practice has been eliminated in some companies.

3. *Copy:*, followed by a list of other recipients of the letter. List these names at the end of the letter. An older notation, *cc:*, is still used by many businesses. At one time *cc:* meant "carbon copies." Today *cc:* is used to refer to copies, just as *pp* is used to refer to pages.
4. The *PS* or postscript. This is used sparingly, and is found most often in sales letters. Technical writers rarely use a postscript, partly because word processors make it easy to add afterthoughts within the text, but mainly because a postscript establishes too informal a tone for a serious technical or business document.

Continuation Pages When your letter continues to a second page, it is necessary to head the page. This heading keeps pages consecutively numbered for reattachment should they become separated.

EXAMPLE

Carter Eggley	2	September 22, 1987
(name of recipient)	(page number)	(date of letter)

Type continuation pages on plain sheets of paper, not letterhead paper. Normally, continuation pages occur in reports, which tend to be longer than general business correspondence. Head all pages after the first one in the same manner. The continuation page heading may vary slightly from company to company, but the form given here is a widely used one.

FIGURE 31-1 A BLOCK-STYLE LETTER

300 West Mineral Street
Platteville, WI 53818
September 22, 1990

Mr. Carter Eggley, Manager
March Enterprises
3725 Locust Street
Rockford, IL 61108-0004

Dear Mr. Eggley:

I am submitting the enclosed proposal, *Methods for Evaluating the Safety Controls in Plant 15-C.* Notice the emphasis on new methods for dealing with the problems prevailing in the plant.

For example, the section on the robots on assembly line 622 offers a new and economical way to implement worker protection.

If this proposal meets with the approval of your board, I will be happy to meet with you to arrange the details for my complete study.

Please call me if you have any questions.

Sincerely yours,

Ellen Rumsford

Ellen Rumsford

encl.
ER, SN

FIGURE 31-2 A SEMI-BLOCK-STYLE LETTER

300 West Mineral Street
Platteville, WI 53818
September 22, 1990

Mr. Carter Eggley, Manager
March Enterprises
3725 Locust Street
Rockford, IL 61108-0004

Dear Mr. Eggley:

I am submitting the enclosed proposal, *Methods for Evaluating the Safety Controls in Plant 15-C.* Notice the emphasis on new methods for dealing with the problems prevailing in the plant.

For example, the section on the robots on assembly line 622 offers a new and economical way to implement worker protection.

If this proposal meets with the approval of your board, I will be happy to meet with you to arrange the details for my complete study.

Please call me if you have any questions.

Sincerely yours,

Ellen Rumsford

Ellen Rumsford

encl.
copy: David Adams

31.2 ESTABLISH A READABLE STYLE FOR BUSINESS LETTERS.

The style of your letters is extremely important. They should not be stuffy, overwritten documents. Rather, they should be friendly, conversational, and clear.

a. Avoid pompous or empty clichés.

The following is a list of clichés or overblown phrases that frequently appear in correspondence. Opposite these clichés are improved versions.

Weak	Improved
Enclosed please find	I have enclosed
Attached please find	I have attached
I am forwarding herewith	I am forwarding
I am in receipt of your letter dated June 22.	I received your June 22 letter.
I beg to differ with your statement of June 3.	I disagree with your June 3 statement.
I humbly request that you consider my application.	Please consider by application.
I beg to acknowledge receipt of your report.	I received your report.
I wish to express my humble gratitude.	Thank you.
We are in hopes that you succeed in your new employment endeavor.	Good luck with the new job!
Please be kind enough to grant me an interview.	May I have an interview?
I am awaiting your earliest reply.	I hope to hear from you soon.
I regret to advise you that I cannot honor your request.	We cannot fill your order.

Per your inquiry of	You inquired
In reply to your letter of August 22 in which you stated . . .	In your August 22 letter you stated
If you will kindly inform us	Let us know
In accordance with your wishes	As you requested
In view of the above	Because
As per your request	You requested
Having received your letter, we	We received your letter.
It is imperative that you contact us at your earliest possible convenience.	Please contact us as soon as possible
Please be advised that my new address is . . .	My new address is . . .
This writer	I
At the present time	now
At this point in time	now
In the immediate future	Soon

b. Establish a "you" attitude.

Always consider your reader's feelings and reactions. Establish a positive attitude and address the reader as "you" whenever possible, avoiding "I" or "we."

EXAMPLES

"I" Attitude	Improved "You" Attitude
I am requesting that I receive a replacement immediately.	Please send me a replacement as soon as you can.
I can come for an interview when I am not in class.	Perhaps you can arrange an interview at a time convenient for both of us.
Please note that I have written	You have heard from me before

about this matter on two other
occasions.

about the confusion in names.

c. Make the purpose clear as quickly as possible.

It is wise letter-writing to state the purpose of the letter in the opening paragraph. The rest of your letter is then developed as specifically and briefly as possible. Clarity and politeness are keys to effective letter-writing.

d. Be brief.

Business and technical people have many pieces of correspondence cross their desks each day. The shorter your message, the more likely it will be read. To say something on one page is to get it read. Letters of two or more pages may never be read unless they are reports busy executives can skim. Long letters usually employ headings for easier, more rapid reading (see Section 19).

The following work sheet will help you organize the body of a letter.

Directions: Fill in the answers to the following questions.

Prewriting

Who is my reader?
What will this reader need to know?
What is my purpose for writing this letter?
What facts do I need to present?
How will my reader react to what I have to say?

Editing

What is my purpose?
Do I state my purpose clearly in my opening paragraph?
Do I use any overblown phrases that obscure my meaning?
Am I specific?
Am I polite?
At all times do I try for brevity in a plain style?

31.3 ADDRESS BUSINESS ENVELOPES CORRECTLY.

Business envelopes come in two standard sizes: 4 by 10 inches and 3½ by 6½ inches. The format for either size is the same. Two items of information appear on the envelope: your return address and the name and address of the person to whom your letter is addressed (the inside address). Type this information neatly and correctly on the envelope, as shown in Figure 31-3.

In the upper left-hand corner, your name and address should match the return address in your letter. The state can be abbreviated in both addresses if proper US Post Office abbreviations are used (see Section 13.6, p. 93, for a complete list of these abbreviations). Center on the envelope the name and address of the person who will receive your letter and make sure it is identical to the inside address.

FIGURE 31-3 PROPERLY ADDRESSED ENVELOPE

Ellen Rumsford
300 West Mineral Street
Platteville, WI 53818

Mr. Carter Eggley, Manager
March Enterprises
3725 Locust Street
Rockford, IL 61108-0004

●**EXERCISE 1** Write a letter to your instructor outlining your expectations for the course. Use either the block or the semi-block format and try to employ the "you" attitude as much as possible.

●**EXERCISE 2** Save all the letters you receive in the mail for a two-week period. Compare both their style and format.

●32

MEMOS

KEY POINTS

- Use a memo only within a company or group as a written form of internal communication. Do not send a memo to a customer.
- At the top of the memo always include the name of the person or group receiving the memo, the name of the sender, the date, and a subject heading.
- Employ an informal tone.

Memos are used within a company or an organization for transmitting information from one person or group to another person or group. They contain a variety of information, but mainly they are used for requests and responses, reports, and directives.

32.1 USE AN ACCEPTED FORM.

Many companies print their own memo forms. Whether the form is printed or has to be typed on plain or letterhead stationery, the four top lines are usually the same. They are in the following format:

TO:
FROM:
DATE:
SUBJECT:

Note: The order in which the four top lines appear varies. They do not have to appear in the order given above.

In the format given, each item of information is on a separate line, with the subject of the memo last, stated in a short phrase.

32.2

The notation *Copy:* or *cc:* often appears on a memo with a vertical list of names after it. See Section 31.1, p. 189, for a complete discussion of the use of *cc:* or *Copy.* Normally, you place this list at the end of a memo, although a few companies prefer it at the top in a column parallel with the four major headings.

Most people sign their initials next to their names in the memo headings. Others prefer to sign their full names either by their typed names or at the bottom of the memos. Whichever method you choose is acceptable as long as you do not insert a complimentary close at the end of your memo. Figure 32-1 shows a sample memo.

32.2 USE AN INFORMAL TONE.

Because memos go to other members of the same group, company, or organization that you belong to, address the reader in an informal tone. In general, make your memos brief, clear, and informal, although your report memos usually will be longer, depending on the quantity of information they must deliver. Write your memos much like the body of a letter. See Section 31.2, p. 195, for a discussion of letter style.

FIGURE 32-1   SAMPLE MEMO

MARCH ENTERPRISES

INTER-OFFICE CORRESPONDENCE

TO: Bud Petri

FROM: Carter Eggley   *CE*

DATE: October 2, 1987

SUBJECT: Rumsford Proposal

I am forwarding *Methods for Evaluating the Safety Control Plant 15C.* It appears to have a number of merits. Read it and send it on to Doug Flannigan and then we can call a meeting to decide if we should accept it.

Rumsford has done some excellent studies for us in the past, but we need to discuss whether we really need this one.

●**EXERCISE 1**   Write a memo to one of your professors outlining a project you are going to do for the course. Compare this memo to the letter you wrote to your professor in Exercise 1, Section 31. In what ways are the styles of the two documents the same? In what ways do they differ?

●33

CORRESPONDENCE FOR JOB-HUNTING

KEY POINTS

- Use common sense in organizing material for a resume.
- Employ action verbs throughout the resume.
- Be sure the resume is easy to read.
- Include only pertinent information on the resume.
- Personalize each letter of application.
- Write follow-up letters as carefully as the letter of application and resume were written.
- Carefully proofread all resumes and letters for typos and spelling errors.

One of the most important letters a person ever writes is a letter of application. Because this letter and its accompanying resume are the job searcher's chief tools for competing in today's "buyer's market," the job applicant has to make the most of them. Although prospective employers read the letter of application first, most people write the resume first because it describes their qualifications in more general terms, while the letter of application specifically addresses a company.

33.1 DEVELOP A CONCISE WELL-WRITTEN RESUME.

a. Make an inventory of all your qualifications and experiences for the resume.

When writing a resume, first make an inventory of your abilities, for, above all, a resume is your personal list of accomplishments that attempts to sell your marketable qualifications to an employer. These qualifications should include *Education, Abilities, Work Experiences,* and *Activities.*

The *Education* portion includes all colleges you have attended and any degrees you have received, along with the dates you received them, listed in *reverse* chronological order, beginning with the most recent degree. If you have earned a bachelor's degree from a college or university, you no longer need to include a high school diploma in this section.

Additional information about your college career is usually unnecessary in this section because the *Activities* section at the end of your resume is reserved for college as well as for community activities.

Too often it is difficult to give a prospective employer a true picture of what you are like. You know you have a great deal to offer a company. You are a capable, hard-working person, but you do not know how to communicate this information. A list of the *Abilities* or *Capabilities* that you feel are some of your strongest assets can give an employer a more rounded picture of you. These abilities can include knowledge of other languages, special technical skills, and personal qualities.

Work Experience is a very important part of your resume. As you did with *Education,* list *Work Experience* in reverse chronological order, beginning with your most recent experience. It is important that you account for every year—no time periods should be left to the reader's imagination. Even very menial jobs show your willingness to tackle a job. By using active verbs that express clear achievements, you can enhance the responsibilities and importance of a job. A list of useful verbs follows.

ACTION VERBS FOR RESUMES

accelerated	developed	motivated	revised
adapted	devised	originated	saved
administered	directed	participated	scheduled
analyzed	established	performed	serviced
approved	expanded	planned	set up
commanded	generated	presented	solved

completed	guided	programmed	streamlined
conducted	implemented	proposed	strengthened
contributed	improved	provided	structured
controlled	increased	ran	succeeded
coordinated	influenced	recommended	supervised
created	inspected	reduced	surveyed
delegated	interpreted	reinforced	tested
demonstrated	introduced	reorganized	trained
designed	maintained	reviewed	transformed

Note: If you have work experience related to the job you are seeking, place it in a section titled *Related Work Experience.* You then can place other non-job-related experience under *Other Work Experience.* Once you are established in the business world, you need to include on the resume only those jobs held since college graduation.

Activities are optional on a resume. When given, they demonstrate that you have been involved in more than your studies or that you are willing to become involved with company and community activities. These activities should be meaningful community and college activities, not just a list of your hobbies and interests unless your hobbies are job-related. Then you should place them in the *Abilities* section.

b. Organize the resume in a clear and easily read format.

Neatly type or print your resume on good quality $8\frac{1}{2}$- by 11-inch bond paper or, if you prefer, have your resume printed. Be sure to look at examples of the printer's work before you contract the job. If you own a computer or have access to one, remember that many printers are now letter quality or near letter quality, so you may be able to print all your resumes using your home computer.

Arrange your material carefully to promote readability. Leave plenty of white space around individual entries and use boldface type or underlining or both to help emphasize individual sections of the resume.

After your prototype resume is finished, carefully proofread it for typographical errors and spelling. The final draft must have neither. Finally, your resume can be photocopied on good bond paper if it was typed, run off on bond paper on a printer, or mass produced at a printer's shop. High-quality typing or printing and paper are a must.

Note: Be sure to proofread the first copy a printer sets up. You do not want to be so unfortunate as to be the unlucky owner of 100 resumes that have glaring typographical errors or misspellings in them. Remember that you are responsible for the errors.

Try to get a resume on one page if possible; however, if you have valuable experience, a second page is sometimes necessary. This is especially true for people who have several years of experience and are looking for a new job rather than a first job. Normally, a resume is *no more* than three pages long. There is a limit to what an employer has time to read!

Figures 33-1 and 33-2, pp. 206–09, give examples of two very different yet readable, clear resumes. They are for the same student but each makes the most of different experiences and of format preferences.

c. Do not clutter a resume with unnecessary information.

Many resume writers consider personal data important, but much of what is labelled as *Personal* is not needed in resumes. For example, unless you are applying for a job as a police officer or for some other job where physical size might come into question, it is a waste of space to give your height and weight. It is equally unnecessary to include the color of your eyes and hair, your social security number, college courses taken, unless relevant and exceptional, and personal hobbies such as reading and watching sporting events. It is illegal for a prospective employer to ask about your marital status, health, nationality, or religion.

Some people prefer to include a "career objective" in their resumes. If you do so, tailor the statement to each company you apply to and include it in the letter of application, not as an overly general statement at the beginning of the resume.

33.2 WRITE A CORRECT LETTER OF APPLICATION.

a. The letter of application should be personalized for each potential employer.

Do not use a form letter as a letter of application. Even though you should develop a basic model, each letter should state something about the company being applied to or about the way in which you learned about a possible job opening with the company. For example, you may have read about specific work the company is doing or you may have heard about the company through a friend, a professor, or an advertisement. If you know someone who works for the company and have permission, be sure to include this, preferably in your opening paragraph.

The impression your letter will make is important, for the prospective employer reads it before the resume. If your letter is incompetently written, the employer may never read your resume, for he or she will put it aside in favor of other letters that are better. It is therefore imperative that your letter makes a good impression.

b. Include all necessary information in the letter.

Begin with a self introduction. In it state exactly what type of position you are seeking. If you are a recent college graduate, include this in the opening paragraph, too. Finally, mention something about the company that relates to your job search. For example, tell the company why you want to work for *them* in particular.

FIGURE 33-1   STUDENT RESUME

JANE F. HASSEMER

SCHOOL ADDRESS (until 5/9/87)   **PERMANENT ADDRESS**
   30 So. Chestnut St., Apt. 7   10713 N. Diamond Rd.
   Platteville, WI 53818   Whitelaw, WI 54247
   (608) 348-5514   (414) 682-3170

EDUCATION
University of Wisconsin-Platteville, B.S., May 1987
Major: Civil Engineering
Emphasis: Transportation Engineering

WORK EXPERIENCE

Westbrook Associates, Inc., Plain, WI
                          January 1986–August 1986
Engineering Technician (Co-op Program)
   Performed all elements of roadway/bridge project design, in-

---

cluding surveying, hydrologic analysis, bridge sizing, roadway alignment, hydrologic report preparation, and plan and public display preparation and presentation. Coordinated with field crews and computed elevations for large bridge projects in Kansas City. Utilized "Hydraulics of Bridge and Culvert Waterways" and "LOTUS" programs.

State of Wisconsin-Department of Transportation, Green Bay, WI

May 1985–August 1985

Student Engineer Trainee—Construction

Inspected materials during the construction of bridge and paving projects. Inspected and controlled work of four-man pipe crews. Surveyed, staked, and made calculations for roadway and bridge projects. Developed drawings for field revisions.

May 1984–August 1984

Student Engineer Trainee—Design

Designed alternates for substandard roadways. Developed quantity and cost estimates for projects. Prepared plans and displays for public presentations.

May 1983–August 1983

Student Engineer Trainee—Construction

Completed preliminary surveying and staking for construction of a new highway. Inspected work of pipe crews and ensured quality control. Recorded daily project progress in diaries.

UW-Platteville Athletic Department, Platteville, WI

January 1985–present

Recruit and Camp Processor

Compiled recruit and camp information on a computer. Coordinated with football coaches to obtain computer scouting reports.

HONORS AND ACTIVITIES

Chancellor's List

Dean's List

Chi Epsilon-National Civil Engineering Honor Society (President, Treasurer, Secretary)

Society of Women Engineers (President, Vice-President)

American Society of Civil Engineers Member

Engineering Advisory Council Representative

Accepted to Who's Who Among Students in American Universities
& Colleges
Recipients of several engineering scholarships
Volunteer for handicapped swim program
Participant in intramural sports (volleyball, softball)

REFERENCES
Available on request.

FIGURE 33-2 STUDENT RESUME

JANE HASSEMER

HOME ADDRESS
10713 N. Diamond Rd.
Whitelaw, WI 54247
(414) 682-3170

CAMPUS ADDRESS
30 So. Chestnut St., Apt. 7
Platteville, WI 53818
(608) 348-5514

EDUCATION
May 1987   B.S., Civil Engineering
University of Wisconsin—Platteville

CAPABILITIES
* Provide ideas based on previous experience.
* Compatible with associates and coworkers.
* Handle projects as an individual or in conjunction with others.
* Adapt to new processes and environments.

EXPERIENCE
* Designed roadway projects and coordinated displays for public
  meetings.
* Presented project proposals on individual basis to highway com-
  mittee, general public, and local government officials.
* Developed cost estimates for projects.
* Performed hydrological analyses for structure sizing and loca-
  tion by means of HEC/2 computer program.
* Analyzed bridges for the determination of the size and location of
  jacks required to raise structures.

* Coordinated with survey crews from Kansas City projects and computed their information necessary for structure layout in the field.
* Inspected and controlled the work of pipe crews.
* Tested concrete to assure quality control in the construction of structures and placement of pavement.
* Surveyed and staked several road and bridge projects.

January 1986–August 1986	Westbrook Associates, Inc.
	Plain, WI
	Engineering Technician
Summers 1983, 1984, 1985	State of Wisconsin
	Department of Transportation
	Student Engineer Trainee
	(Construction and Design)
January 1985–Present	UW-Platteville Maintenance Dept.
	Computer Operator

HONORS AND ACTIVITIES
Chancellor's list
Dean's list
Chi Epsilon-National Civil Engineering Honor Society (President, Treasurer, Secretary)
Society of Women Engineers (President, Vice-President)
American Society of Civil Engineers member
Engineering Advisory Council representative
Who's Who Among Students in American Universities & Colleges
Recipient of several engineering scholarships
Volunteer for handicapped swim program
Participant in intramural sports (volleyball, softball)

In the body of the letter, list your qualifications for the job you are applying for. You may have other qualifications that you feel are more impressive, but if they do not relate to the job you are applying for, they should not be mentioned because they already appear in your resume. The body of your letter also should state that you have enclosed your resume.

The conclusion should clarify when you are available for a job and how you can be reached. It is important to note that you should

33.2

never end your letter with "thank you." No one has done anything for you yet. A statement such as "I look forward to hearing from you" is appropriate.

Usually you conclude the letter with *Sincerely yours,* although a number of young people prefer to use *Respectfully yours,* possibly to display respect for elders.

Below your typed signature, type *enc., encl.,* or *enclosure* flush with the left margin to indicate that your resume is attached. Staple your resume if it is more than one page in length; however, *never* staple together your resume and letter of application.

Typist's initials and a *PS* (postscript) should *never* appear on a letter of application. Your letter is assumed to be your work. Figure 33-3 is a sample student letter.

Finally, your letter should be neatly typed on the same kind of paper as your resume. Carefully proofread it for any typographical or spelling errors and neatly correct them. If the letter needs a number of corrections, you should retype it. If you are using a word processor, make the changes on your disk and run a corrected version of your letter.

FIGURE 33-3 STUDENT LETTER OF INTRODUCTION

30 So. Chestnut St., Apt. 7
Platteville, WI 53818
February 12, 1987

Robert Darr
Coe and VanLoo
4550 N. 12th St.
Phoenix, AZ 85014

Dear Mr. Darr:

After speaking with both you and Joe Marcuson of JHK and Associates, I became interested in the work in which Coe and VanLoo is involved. I am seeking a position as a Design Engineer in the transportation-related areas. I will graduate from the University of Wisconsin-Platteville in May, with a Bachelor of Science degree in Civil Engineering and an emphasis in Transportation Engineering.

Through my internship and summer jobs, I have acquired a considerable amount of experience in both design and in the field. These opportunities have given me a good understanding of the coordination between design and actual construction, and I feel this would be beneficial in my employment with Coe and VanLoo.

Enclosed is a resume that gives further information about my credentials.

I will be in Phoenix March 14 through March 21, and I'd appreciate the opportunity to interview with you at your convenience. I can be reached at (608) 348-5514, or a message can be left at (608) 342-1567. If I have not heard from you by March 4, I will contact you then to discuss the possibility of an interview. I am looking forward to hearing from you.

Sincerely,

Jane F. Hassemer

Jane F. Hassemer

Enc.

33.3   WRITE FOLLOW-UP LETTERS NEATLY AND CORDIALLY.

a. Write a thank-you letter after each job interview.

A follow-up thank-you letter is customary and polite. Wait for a few days and then send a letter expressing an interest in the position and a thank you for the interview. The following is an example of the body of a thank-you letter:

The trip to Apex was unique. I never realized how many steps went into packaging your product. As I mentioned at my interview, I have had three summers' experience working on an assembly line. This experience, together with my degree in Industrial Engineering, should make it easy for me to adapt to assembly-line studies for your company.

Thank you for your time and courtesy. After touring your plant, I am excited about living in Carbo and being a member of your team.

b. Carefully write acceptance letters.

In a letter of acceptance, remember two important things:

1. to be polite, and
2. to carefully and clearly spell out the terms of the offer. The following is an example of the body of an acceptance letter.

I am happy to accept an entry-level position with Vanity Cosmetics as an industrial engineer in the mascara and eye-shadow division, with a starting salary of $22,500.

I will be ready to work on July 17, 1988, when I report to Amelia Makepeace in the personnel office.

I look forward to working with your growing company.

c. Employ a friendly attitude in letters of refusal.

Letters of refusal are more difficult to write and occur less frequently for most job seekers than do thank-you letters. You want to maintain friendly terms with the company and do so by writing a polite letter. In the future, the company you turned down may be the very company you want to work for. The following is an example of the body of a letter of refusal.

I was highly impressed with Apex. When I received a job offer from you I was flattered that you considered me worthy of your company. Unfortunately, at this time I am unable to accept your offer. Instead, I have taken a job with Vanity Cosmetics in Joseph, Pennsylvania, which is close to a major university where I can pursue an MBA in the evenings. Thank you for your time and confidence.

33.3

●**EXERCISE 1**   Find a book in the library on resume writing. Compare the different forms for a resume. What do they have in common? How do they differ? Are all the resumes written for the same type of job search? For the same type of applicant?

●**EXERCISE 2**   Explain the importance of the letter of application in the job search.

●**EXERCISE 3**   Write a letter of application and a resume for either a summer job or for an internship.

●34

ABSTRACTS AND SUMMARIES

KEY POINTS

- In general, abstracts are briefer than summaries.
- Abstracts mirror the long paper's organization.
- List the paper's topics in the descriptive abstract.
- Tell the results and conclusions of the report in an informative abstract.
- Use a nontechnical style in executive summaries.
- Remember that executive summaries should make reading the longer report optional.
- Reiterate the report's principle ideas in an end summary, but never add new information.

Abstracts and summaries are both condensed versions of longer pieces of writing and should always be written after the longer report is finished. Busy people use abstracts and summaries extensively; executives ask for summaries of a variety of important material, and professionals from every walk of life rely on the abstracts published in indexes to keep up on the latest research in their field and to find quickly the document with the information they need. For the technical communicator, perhaps the most important abstracts and summaries are those for technical reports.

Abstracts are generally recognized as having two forms: the descriptive and the informative. The former merely lists the topics of the full document; the latter presents in a very condensed style all the major aspects of the original document. Although very brief (often restricted to two hundred words, or 5 to 10 percent of the length of the original document), the abstract mirrors the arrangement of the longer report. It may precede a technical document, giving the reader, usually management, an orderly overview of the report; it may accompany a technical or scientific publication or paper presentation and be printed separately in an index or other source that alerts researchers to the longer report's existence; or it may be incidentally submitted to a busy executive who needs to know the essential information contained in a newspaper or periodical article.

Summaries are also recognized as having two forms, the executive or introductory and the concluding. The former is written in a nontechnical style for fast reading, and it discusses briefly the major aspects of the full document; the latter is a more technical discussion of the body of a report that capsules and reviews the major facets of the full document. The executive summary of a technical report is usually longer than the abstract because the summary is more thorough in emphasizing the results, conclusions, and recommendations. The summary is often restricted to five hundred words in comparison with the abstract's two-hundred-word limit. Sometimes, however, the summary is simply described as a condensation without a specific length being given. It does not necessarily follow the same arrangement as the longer document in the way the abstract does, but attempts to make the report's important points quickly understandable. The concluding summary ends rather than precedes a longer report and is usually shorter than the executive summary, although it has the same emphasis on the major points in the document.

34.1   USE THE DESCRIPTIVE ABSTRACT TO TELL ONLY WHAT TOPICS THE LONGER REPORT CONTAINS.

The descriptive abstract is a statement of your paper's scope or table of contents and its organization. It does not reveal data, results, conclusions, or recommendations. Use the descriptive abstract in indexes or listings to help a reader locate a report that deals with certain topics. You can also help a reader decide whether to read your entire report or certain portions of it. Your descriptive abstract is very brief, usually about three sentences. This kind of abstract is being used less and less as busy people want to have more informative abstracts that can save them the trouble of reading the document.

34.2

EXAMPLES

The New Uses for Long-Wavelength X-rays
by Ronald Holsten

The x-rays of radiography and crystallography are short-wavelength. The long-wavelength x-rays lie between ultraviolet radiation and short-wavelength x-rays. The use of long-wavelength x-rays in microscopy, astronomy, and microelectronics may be possible.

An Examination of Wilderness Programs in Small Colleges
in the Tri-County Area
by Janet Schollanden

This report examines which type of wilderness program—organization, management, or adventure—small colleges in northeast Schaun, southwest Dareen, and southeast Brown counties have employed. Some colleges have changed their wilderness program choice one or more times. The success of the programs, past and present, is determined by trends in enrollment and retention.

Note: The descriptive abstract may or may not refer directly to the report. The use of such phrases as "this report" is sometimes restricted by the employer or publisher. If you are forbidden to make reference to your report, style your sentences like the following revision of the preceding example: "An investigation into the types of wilderness programs offered by colleges in this and surrounding counties revealed that many have tried several of the types in a search to find the one most effective in their area."

34.2 USE THE INFORMATIVE ABSTRACT TO PRESENT THE MAJOR CONTENT AREAS AND CONCLUSIONS OF THE LONGER DOCUMENT.

The informative abstract tells the subjects, methods, results, conclusions, and recommendations of the report while considerably condensing the material through the omission of data and other details. A general rule is that the abstract is not to exceed 10 percent of the full document's length and is usually shorter. In the informative abstract, retain the report's arrangement and perspective. Executives and other busy people frequently read the informative abstract instead of the report, whether it is found at the beginning of a long document or is independent of the report. At all times, give the reader enough information to decide whether to read the document.

EXAMPLES

An Investigation into Feasible Criteria for Selecting Geometrically Designed Quilts to Be Exhibited
by Joan Czymanski

Groups that choose to have a geometrically designed quilt exhibition must decide how to choose the works for display. Twenty associations statewide that had held quilt-design shows were surveyed as to the criteria of selection, the success of the show, and the opinions of the exhibition directors about the criteria used. Related documents published by the National Quilt Design Association (NQDA) were consulted, and administrators of NQDA were interviewed. The most successful shows had selected the exhibitions according to originality, execution, and aesthetics. The work should incorporate technology and humanity and the medium should be a hand-done, finished, two-dimensional work of art using any materials.

Alternative Internship Procedures for Use in Everon's Management Training Program
by Francesca DeManna

Everon Inc. of DesPlains, Wyoming, which has no formal training program for new management personnel, investigated the potential results of initiating one of two possible internship procedures: extensive use of mentors within the company; and use of a corps of hired professional management training experts from Maximum Management Training Company (MMTC). Thirteen firms statewide that had used MMTC were interviewed, as were three local firms who had used mentoring programs. Information obtained directly from MMTC and two similar firms also proved helpful. Potential mentors within

Everon returned completed survey forms. Areas of consideration included cost, space, personnel use, compatibility. The following conclusions were reached: MMTC would cost three times more than would the in-house mentoring procedure; new management personnel would have to leave the premises for the training period under MMTC; mentors could only give a total of 50 hours for any new trainee; company procedures would be clearer if in-house mentors are used. The recommendation is that Everon initiate a formal training procedure for new managers, using an in-house mentoring program.

34.3 PREFACE A LONG TECHNICAL REPORT WITH AN EXECUTIVE SUMMARY.

The executive summary condenses the long report that it prefaces, but presents the report's material completely enough to make reading the report optional. The executive summary is usually written for readers who are not technically oriented. As a summary, it is longer than an abstract and more complete. The length of the summary varies with the length of the report. Many times the summary is only one or two pages long. However, for a long report, the summary can run twenty-five pages.

The executive summary has two parts: a condensation of the report's introduction—background and statement of the problem as well as scope and purpose of the report, and a brief description of the major facts from the body of the report with conclusions and recommendations.

Because of their importance, the summary's conclusions and recommendations are usually given special subheadings. Material is not necessarily ordered in the summary as it is in the report, but instead concentrates on making clear the results, conclusions, and recommendations, so that the reader can make sound decisions based on the information.

EXAMPLES

ALTERNATIVE INTERNSHIP PROCEDURES FOR USE IN EVERON'S MANAGEMENT TRAINING PROGRAM

by Francesca DeManna

Everon Inc. of DesPlains, Wyoming, has had no formal training program for new management personnel, but rather a two-day preemployment period when new managers intensely observed company operations. Everon investigated the potential results of initiating one of two possible internship procedures for training new management personnel. One procedure involves extensive use of mentors within the company; the other makes use of a corps of hired professional management training experts from Maximum Management Training Company (MMTC).

Thirteen firms within the state were found to have used MMTC; executives responsible for hiring the professionals, as well as managers who had experienced the procedure, were interviewed. Information obtained directly from MMTC and two similar firms also proved helpful. Potential mentors within Everon contributed their views on feasibility and effectiveness through written surveys. Executives from three local firms that use in-house mentoring procedures also were interviewed. Areas of consideration included cost, space, personnel use, and compatibility.

Conclusions

1. MMTC would cost three times more than would the in-house mentoring procedure.
2. The available space could not be kept available eight hours a day for two weeks, so new management personnel would have to leave the premises for the training period.
3. Mentors for in-house training procedures could only give a total of 50 hours for any new trainee.
4. New management trained by company mentors would know company procedures much more than if trained by a professional group.

Recommendation

Everon should initiate a formal training procedure for new managers, using an in-house mentoring program.

REPLACEMENT FOR THE MAINFRAME AT SYCAMORE STATE UNIVERSITY

Background

The mainframe at Sycamore State University is outdated and must be replaced. The university's financial officer has done a preliminary investigation to determine that the choice is between a system of 30 networked PCs versus the purchase of a Megacrunch 999 mainframe and 20 workstations. The university already has 10 workstations.

Methods

Areas of investigation include capital investment costs and average time of running certain benchmark programs. It was assumed that all terminals will be in use when programs are run. Investigators spoke with administrators and faculty from Buckson State College, which has a 100 PC network, and with Gerisonne, a local company that owns a Megacrunch 999. Sales brochures and evaluative periodical articles were also checked.

Conclusions

1. The capital investment will be 10% less for the PCs, but future expansion will cost 25% less per station for the mainframe.
2. The running time for simple programs, such as Super Spreadsheet A, B, and C, showed the PCs' operating time as 2% less.
3. The running time for complicated programs, such as Monaco 100, showed the mainframe's operating time as 7% less.

Recommendation

The Megacrunch 999 should be purchased because expansion costs and running time for complicated programs will be less. The PCs may be unable to handle the even more complicated programs that will arise in the future.

34.4 USE A SUMMARY IN THE CONCLUSION OF A REPORT THAT NEEDS TO HAVE ITS FACTUAL INFORMATION RESTATED.

Many reports are too short or simple, argumentative, or descriptive to need concluding summaries. When your report presents a body of statistics or facts, you may not need one. If you do, your summary should briefly restate the document's purpose. It then should review, without elaboration, the principle points made in your discussion. Present this review in paragraph form or in some kind of enumerated list (see Section 30) at the end of the report on the collected data. Conclusions and recommendations may or may not follow. Do not use new ideas or information in this review. In addition, list principle points in the same order they appear in the body of the report.

EXAMPLE

SUMMARY

All of the instructors in the city schools must be evaluated by an assessment tool that philosophically corresponds to the Wyrick Teaching Model adopted two years ago. The assessment tool now being used to evaluate the media specialists must be replaced by either the Critical Approach Evaluation (CAE) or the Objective Approach Evaluation (OAE). The CAE uses defined factors to grade the job against an established scale. The OAE involves the supervisor and instructor working together to establish goals and to determine whether the goals have been reached, or at least furthered. Although the CAE has been proved effective in evaluation as well as instructor motivation, it fits less well than OAE into the philosophies of the Wyrick Teaching Model. The OAE is a newer evaluation tool, however, and may yet prove as effective as the CAE.

●**EXERCISE 1**   Your instructor may have a formal report, the body of which can be distributed to class members. Read carefully the body of the report, then write a descriptive and an informative abstract, and an executive and a concluding summary. After you have written yours, it would be beneficial to compare your writings with those of others in the class as well as with that written by the report's author.

●**EXERCISE 2**   Find a technical or scientific report in a journal or magazine, and write an informative abstract of it.

●**EXERCISE 3**   Find the formal report in this text (Section 36) or in some other text that does not have an executive summary with it and write one for it.

34.4

●**EXERCISE 4** If the class composes formal reports, exchange the body of your report for the body of someone else's report. Write a descriptive and/or informative abstract and an executive and a concluding summary for the other person's report. Then compare yours with the one the author of the report wrote. The author may thus become aware of any clarity and emphasis problems in the document.

●35

INFORMAL REPORTS

KEY POINTS

- Write informal reports in a style appropriate for the intended readers.
- Give informal reports a good appearance.
- Write informal reports in an informal style where possible.
- Check the accuracy of all information in an informal report.
- Write informal reports clearly and concisely.

The most common classification of technical reports is into two types: formal and informal. The characteristics that distinguish them most from each other are form and content. Informal reports are usually shorter than formal reports and are often sent as letters or memos. Formal reports are usually more comprehensive and consist of several sections or chapters. Formal reports are written in a formal style while informal reports may be written in either formal or informal style (see Section 21).

35.1 CHOOSE A STYLE OF INFORMAL REPORT THAT IS APPROPRIATE FOR THE INTENDED READER.

The purpose of a report is to convey information effectively to the reader. Your choice of writing style in informal reports depends mainly on the relationship between you and your intended reader.

Your style is also influenced by the subject of your report. In general, use an informal literary style to achieve effective communication and a sense of personal contact between you and your reader. For example, if a report contains mostly personal observations, use "I observed" instead of "it was observed."

The format of informal reports varies, depending on your preferences as the writer, the type of report, and the needs of the intended reader of the report. At minimum, your format usually includes an introduction, discussion, and conclusion.

35.2 WRITE INFORMAL REPORTS CAREFULLY.

Informal reports should have the following characteristics: a good appearance, an appropriate style, accurate information, and clear and concise writing.

a. Be sure an informal report has a good appearance.

Since the appearance of your report is the first thing that the reader notices, you need to prepare an attractive first page or cover. The appearance and attention to detail in your report reflect on your credibility and that of the organization you represent. Produce your report on quality paper, employ professional illustrations, and use readable type. Leave an appropriate amount of white space surrounding the written text and illustrations. Arrange the parts of the report to ease the reading and understanding of it.

b. Write an informal report in an informal style wherever possible.

In informal reports your emphasis should be on effective communication. Correctly use grammar and punctuation. Use the active voice whenever possible, but vary sentence structure and length to make the report interesting and flow smoothly.

c. Be sure of the accuracy of the information in the report.

Check the details of the report to eliminate mistakes in the data or graphics, inconsistencies in the nomenclature, and other erroneous information. The credibility of your report is suspect if even seemingly minor errors remain. The misplacement of a decimal point, for example, could result in the loss of a contract.

d. Be sure the written portion of the report is clear and concise.

Present only important information, and use a format that is easily and quickly understood. Include tables and figures when they can best present data clearly. In the prose part of your report, focus on the topic of the report with appropriate emphasis on the main points.

35.3 UNDERSTAND THE VARIOUS TYPES OF INFORMAL REPORTS.

Six types of informal reports are considered in this section: inspection reports, incident reports, progress reports, laboratory reports, periodic reports, and change orders.

Inspection Reports   Inspection reports describe the condition of a piece of equipment, the data examined, or tests performed. The description is based on firsthand knowledge and is often submitted in the form of a letter.

The information in an inspection report usually has the following format: introduction, discussion, conclusions, and recommendations.

In the *introduction*, present the background information about the piece of equipment, building, or other item being inspected. Explain the circumstances and purposes of the inspection.

In the *discussion* section, outline the facts obtained during the investigation and describe the methods used to obtain the facts. Outline alternatives to resolving the problem or improving the situation. Analyze each of these ideas in sufficient detail to communicate clearly what the problems are.

Reach your *conclusions* based on the facts you presented and, when appropriate, make *recommendations*. The conclusions are a summary of the findings of your report. The recommendations are the actions that you feel should be taken as a result of your study. Figure 35-1 is an example of an inspection report.

FIGURE 35-1 INSPECTION REPORT

Carroll D. Besadny
Secretary

State of Wisconsin
DEPARTMENT OF NATURAL RESOURCES
Southern District   3911 Fish Hatchery Road
Fitchburg, Wisconsin   53711-5397

March 31, 1988   3310

Annette Dutcher, City Clerk
City of Platteville
City Hall
Platteville, WI   53818

Dear Ms. Dutcher:

On March 25, 1988, I made an inspection of the Platteville
waterworks system as part of our annual inspection program.
Mr. Mike Willis, waterworks superintendent, and Mr. Dan
Hibner, certified waterworks operator, accompanied me during
the inspection. You might recall that the previous such
inspection was made on June 11, 1987. Based upon my
inspection and a review of our files, the following items need
your attention:

1. Apparently the City is intending to enforce its cross-
   connection control ordinance and Mr. Willis indicated that
   he soon will be hiring someone to carry out the inspection
   program. Apparently, this hiring process should occur in
   the month of April. If that is the case, the Department has
   no problems with the City proceeding immediately to begin
   the inspection program and to begin keeping records of
   those inspections. Remember that the entire system should

be looked at on a 7 to 8-year rotating basis. We will be asking to see the City's record of such inspections during next year's annual inspection.

2. At the same time, the waterworks personnel should also check for the existence of unabandoned private wells. It is my understanding that the City has passed the private-well-abandonment ordinance. If private wells are found, the owner should be informed that they will be required to obtain a private-well permit or the well should be permanently abandoned. Copies of any well-abandonment reports should be submitted to this office once they are completed.

3. The screen over the end of the air-vacuum relief pipe at well No. 2 needs to be replaced.

4. The two hose bibs in the shop area of the main pump station still need to be provided with hose-bib-type vacuum breakers. This problem was pointed out in last year's letter and still has not been corrected.

5. All of the piping and pumping equipment at all of the well houses should be repainted to prevent further corrosion.

6. There are several potential pollution sources located above and near well No. 2 and No. 3, including a salvage yard, gasoline storage tanks, and a pesticide fertilizer facility. The City should closely monitor the activities of these sites since any leakage, spills, or other careless activities could result in the groundwater becoming contaminated and possibly ruining two of the three City wells. Perhaps zoning changes should be considered for that area to better protect the two municipal wells.

A check of our records indicates that the City has done a job in making sure that at least eleven samples are submitted from the distribution system per month for bacteriological analysis. Our records show that only one such sample was missed in October of 1987 when only ten samples were received. Please be sure that in the future the proper number of samples are submitted each month. In addition, it is required to submit at least one sample per quarter from each active well prior to the point of chlorination. Our records show that this was done during 1987. The waterworks is also required to submit at least one sample from the system per month for fluoride analysis. This is to be a portion of a "split" sample, with the other half of the sample being tested for fluoride residual using the City's fluoride testing equipment. Our records show that this was done during 1987 and the residuals were near the desirable level of 1.1 parts per million. Lastly, all of the monthly pumpage reports were submitted during the past year and were completed in excellent detail. For their good work in regard to the sampling program and the submission of monthly pumpage reports, we wish to commend the City waterworks department.

As you can see, the City waterworks appears to be in relatively good condition. However, some improvement could be shown in regard to the cross-connection inspection program and in regard to the enforcement of its private-well-abandonment ordinance. We hope the City views these ordinances seriously and gives the proper attention to make sure these inspections are done. Again, we will be expecting to see considerable progress in improving these areas during the next year. Please give the other items mentioned in this letter your prompt attention. I wish to thank Mr. Willis and Mr. Hibner for their cooperation during the inspection.

Sincerely,

Delbert L. Maag

Delbert L. Maag
Water Supply Unit Supervisor

DLM:cmt
cc: Dodgeville Area Office
    Public Water Supply—WS/2
    Mr. Mike Willis, Certified Waterworks Operator, Platteville,
    WI 53818
    Honorable David Waffle, Mayor of Platteville, Platteville,
    WI 53818

The report's introduction describes the purpose of the inspection, when the inspection took place, who accompanied the inspector and when the previous inspection took place.

The discussion, conclusions, and recommendations are combined in an enumeration of what was found to be deficient. The actions that the city needs to take to correct each problem are identified when the problem is discussed. (Other reports may separate the discussion from the conclusions and recommendations.)

The report concludes with a complimentary comment about the cooperation of the city in completing actions requested in previous inspections.

Incident Reports Prepare incident reports to describe the facts obtained from an investigation of a particular sequence of events or an accident. The format and preparation of an incident report are very similar to those of an investigation report. Figure 35-2 is an example of an incident report. It contains an introduction, discussion of the causes of the problem, and conclusions. The report is written by a professional civil engineer who was hired to investigate the incident, record his observations, and reach a conclusion regarding the flooding problem.

FIGURE 35-2 INCIDENT REPORT

Roger Nelson, PhD, PE
725 Main Street
Platteville, WI   53818
January 2, 1987

Mr. Fred Smith
First Federal Insurance Co.
Hurley, WI   54534

Dear Mr. Smith,

I was asked to investigate the cause of the flooding of the Acme Lumber Company. On December 20, 1986, I investigated the

---

flooding incident at the Acme Lumber Company. The property contains several connection buildings of the Acme Lumber Company, which is no longer in business.

I observed a pool of water in the center of the company lumberyard. The water appeared to be coming to the surface from below ground. I believe this water to be groundwater because it appeared to contain a high iron content. The iron was oxidized upon contact with the air and formed a brownish-red precipitate on the ground surface around the pool and on the bottom of the pool of water in the center of the lumberyard. The water flowed across the yard from the shallow pool in the center of the yard, under the building in the southwest corner of the property, and into a drain on the adjacent property.

I was told the drain was constructed in 1974 or 1975 and had effectively kept the company lumberyard area from being flooded by a spring on the adjacent property. Water did not start to surface inside the lumber company property until a few days after the city wells to the west of the fire station were capped on March 19–20, 1986. The DNR office in LaCrosse has a copy of the complaint about water surfacing in the company lumberyard and added a copy of the memo to the Hurley Water Supply File. A copy of this memo is attached.

There was damage to the footings of the lumberyard buildings, but it was not possible to determine exactly when the damage had occurred. The damage appeared to be due to frost action. Some of the footings had cracked and/or tipped over sideways and were no longer properly supporting the roof and/or walls of the building. The frost action and damage is not likely to have occurred recently. The ground was not frozen near the pool of water in the center of the lumberyard, and several inches of snow covered most of the yard. The ground around the footings was not frozen to a depth great enough to have caused the damage observed.

After viewing the lumberyard, I walked down the alley to the north of the lumberyard toward the fire station and observed the drain mentioned earlier. Water was flowing into the drain from several directions. The drain's lateral pipes did not appear to be connected to the drain and the drain was not functioning as effectively as it might. The largest flow was from a small ditch, which had recently been dug to drain water from the two capped artesian wells on the western edge of the fire station. There was a large unfrozen and bare area north of the drain. It was swelled from the groundwater coming up beneath the sod and was extremely spongy to walk on.

The groundwater in this area is typically around 45 degrees Fahrenheit and is keeping the grass warm enough that it was still growing and as green as in the summer. (See the map on the second page of the attached DNR memo for location.) There was a large, several-feet-deep pool of water around the wells west of the fire station, and water was flowing out of both of them.

Mr. Gary Jones, from the drilling company that capped the wells, indicated that he thought the water was coming from between the inner 8-inch casing and the outer 10-inch casing on one well and through or around the grouting material in the other one. He plans to come back in the spring and grout between the two casings to stop the flow from the wells.

Capping the wells should stop the flooding and prevent further damage to the Acme Lumber Company.

Sincerely yours,

Roger Nelson

Roger Nelson, PhD, PE

encl.

The *introduction* describes the incident, flooding of the Acme Lumber Company, and the date and location of the investigation.

The *discussion* includes a thorough description of the flooding and damage to the lumber company. This part of the report is lengthy, demonstrating that the writer thoroughly inspected the premises.

The *conclusion* describes a solution to the flooding problem. This conclusion is based on the evidence presented in the previous paragraph, information gathered at the scene of the flooding.

Progress Report Describe in progress reports work that has been completed during a specified period of time. Write the report in the form of a letter or memo or present it orally. A typical report contains the following: introduction, work accomplished, schedule status, and future work. Use subheads to divide the information.

In the *introduction* inform the reader of the period of the report, the context of the report, and the status of the project. Make the report informative for the intended reader, but, because of its specialized information, be aware that your report may not be totally understood by readers unfamiliar with the project.

List the *work accomplished* in chronological order or as a series of tasks in a sequence.

Your *schedule status* comes next. If your completion of work is not keeping up with the schedule, explain your reasons for being behind. Always write optimistically, even when things are not going very well.

Follow this section with a *future work* section. In this final section discuss the time of the next report and the prognosis.

Figure 35-3 is an example of a progress report. The report contains the minutes of a construction progress meeting that were prepared by the design engineer for the monthly meeting between the consulting firm that completed the design and the contractor who was building the wastewater treatment plant.

Laboratory Reports Laboratory reports communicate the results of some type of testing. The format of the report is different from most other types of informal reports. It often is only a table or standardized form in which the results of the testing are inserted. Your report should include the following information: how many and when the samples were received, the procedure used, the accuracy and limits of detection of the test procedure, and the results of the test.

FIGURE 35-3   PROGRESS REPORT

CONSTRUCTION PROGRESS MEETING NO. 22
WATER POLLUTION CONTROL FACILITIES
PLATTEVILLE, WISCONSIN

August 28, 1989

Introduction
This report reviews the progress that has been made since our last meeting and contains a schedule for the next three weeks for the contractor (Acme Construction) and the subcontractor (Shot Electrical).

Acme progress since previous meeting
1. Working on finish grading, curb, gutter, and sidewalks.
2. Working on installation of handrail and grating at final clarifiers and aeration tank.
3. Installed grout on final clarifier No. 3.
4. Painting on final clarifiers.
5. Completed painting and piping at primary digester.
6. Removed and inspected secondary digester cover.
7. Tested all buried pipe.

Shot progress since previous meeting
1. Completed underground conduit.
2. Completed conduit to fuel pumps.
3. Completed pole bases.

Schedule status
Work on the project is still four weeks behind schedule because of the problems encountered in blasting the rock to complete the tunnel between the filter building and the main control building during the extremely cold winter of 1985.

<u>Schedule for the next three weeks</u>
Acme Construction
1. Install replacement parts on final clarifiers No. 1 and No. 2.
2. Complete painting on final clarifiers and R.A.S. pump station and grit/chlorine building.
3. Complete post aeration blowers and chlorination system.
4. Complete curbing and sidewalks.
5. Begin landscaping, paving, and fencing.

Shot Electrical
1. Complete outside lights.
2. Install the dissolved oxygen meters and controls.

<u>Prognosis</u>
The painters have been doing a relatively poor job of painting and covering all sides of exposed steel surfaces in the digester and clarifier. This has resulted in a slower than necessary progress by the painters. If favorable weather conditions continue and we have a warm, dry fall, some of this lost time can be made up. The four weeks that we are behind schedule on the overall project also may be partially made up.

Joe Finn

Joe Finn

Strand Associates

Figure 35-4 is a sample of a standardized form of a laboratory report of some chemical tests. This report consists of three pages of a standard form (only one page is included here) in which the information is presented in a tabular form. The sample type, reference number, type of analysis, amount of each constituent found, limits of detection, and method used in each case are shown. Since the client is only interested in the level found, and numerous tests like this one have been performed for the client, this format is adequate to report the results.

Periodic Reports   A periodic report is issued at regular time intervals and contains information about the status of an organization, project, or the activities of a group. An example of a periodic report is the annual report issued to owners of mutual funds, such as Figure 35-5. A typical periodic report contains the following components: introduction, discussion of the changes in status since the previous report, and supporting information.

A letter from the president introduces the annual report in Figure 35-5. The letter summarizes the economic conditions and expectations for the future. It also summarizes some of the new services provided by the company.

The portfolio manager's perspective indicates the status of the fund since the last report. The perspective also gives a background of the past year's economic factors influencing the fund's performance. Overall, the perspective is positive and is optimistic for shareholders' future investment returns.

The supporting information that makes up the bulk of the twenty pages of this annual report follows the perspective but is omitted here. It contains financial and accounting data about the fund's investments.

Change Orders   Engineers, contractors, and builders use a change order to modify the original contract after a project is partially completed. It may include a change in the equipment specified in the original plans, drawings for construction of a project, tools needed, and amount of work to be performed. The change order *should* include a description of the change, the justification for the change, the cost of the original and the changed item, and the authorization approvals.

Figure 35-6 is an example of a change order. In this change order, the original contract was expanded to include the installation of more water main. The details of the change, 800 lineal feet of 10-inch water main, are indicated along with the reason for the change. The cost of the change and the approval signatures are also included.

FIGURE 35-4 LABORATORY REPORT

A&L ENVIRONMENTAL SERVICES.
A Division of A&L Mid West Agricultural Laboratories, Inc.
13611 'B' Street ● Omaha NE 68144 ● (402) 334-7770

REPORT NUMBER: 6-069-734 Page 1 of 3

PHILLIPS AG. CENTER #1701
ROUTE 4
PLATTEVILLE, WI 53818

LAB NO.	SAMPLE ID.	ANALYSIS
958-1	1-Soil	Total Solids
		Total Volatile Solids
		Ammonia Nitrogen (dry basis)
		Lead (dry basis)
		Zinc (dry basis)
958-2	2	Total Solids
		Total Volatile Solids
		Ammonia Nitrogen (dry basis)
		Lead (dry basis)
		Zinc (dry basis)
958-3	3	Total Solids
		Total Volatile Solids
		Ammonia Nitrogen (dry basis)
		Lead (dry basis)
		Zinc (dry basis)

MARCH 10, 1986

CLIENT: VINCE PHILIPP
         EAST DUB, IL 61025

SUBJECT: ENVIRONMENTAL ANALYSIS

LEVEL FOUND	DET. LIMIT	METHOD
68.67%		Standard Methods 16th Ed.
5.33%		Standard Methods 16th Ed.
100 ppm		Electrode
19.0 ppm	5.0 ppm	Flame AA
91.3 ppm	0.50 ppm	Flame AA
72.32%		Standard Methods 16th Ed.
3.62%		Standard Methods 16th Ed.
38 ppm		Electrode
13.0 ppm	5.0 ppm	Flame AA
70.0 ppm	0.50 ppm	Flame AA
87.28%		Standard Methods 16th Ed.
2.40%		Standard Methods 16th Ed.
34 ppm		Electrode
9.2 ppm	5.0 ppm	Flame AA
60.0 ppm	0.50 ppm	Flame AA

Respectfully submitted,

Cheryl H. Davis

Cheryl Davis
Environmental Services

Dedicated Exclusively to Providing Quality Analytical Services

35.3

FIGURE 35-5 PERIODIC REPORT

Message from the President

Dear Shareholder,

IDS Bond Fund's performance during the past fiscal year reflected the flexible investment strategy of the Fund. The Fund's portfolio manager, Fred Quirsfeld, reviews the performance of the Fund during the past fiscal year and discusses prospects for the coming months in his comments on the following page.

While economic growth continues for the present, IDS expects higher inflation and interest rates to cause a slowdown, perhaps even a recession, within a year or two. A focus on long-term financial objectives and a balanced investment program are appropriate guidelines for investing in this environment.

Bond Fund invests a majority of its assets in high- and medium-grade corporate bonds and may be an appropriate fund for investors who desire a high level of current income with only a moderate degree of risk.

For your convenience, IDS provides several alternatives for your dividend distributions. Most shareholders reinvest their dividends in additional shares of the Fund. But you also may direct them to any other publicly offered IDS fund in which you have an investment. If you need to receive cash dividends, you can sign up to have your dividends sent directly to your bank account. This avoids lost or delayed checks. Contact your IDS personal financial planner for more information on these dividend services.

Your IDS personal financial planner can help you decide how the Fund fits the objectives in your financial plan. If your financial plan calls for an emphasis on other investment and protection objectives, your IDS planner also can tell you about the 26 other publicly offered funds in the IDS MUTUAL FUND GROUP and other IDS products and services. These products and services meet a broad range of investment and protection needs.

Sincerely,

Robert F. Froehlke

Robert F. Froehlke
President
IDS Bond Fund, Inc.
Oct. 7, 1988

From Your Portfolio Manager: A Perspective

Rising interest rates caused bond prices to decline during most of IDS Bond Fund's fiscal year. The exception was the period following the Oct. 19 stock market drop, when fears about a recession caused interest rates to fall and bonds rallied. By taking advantage of the rally and maintaining a defensive posture during the remainder of the year, the Fund was able to reduce the impact of rising rates.

For the 30-day period ended Aug. 31, 1988, the last business day of the year, the Fund's annualized yield was 9.21 percent, based on the maximum public offering price on that day. Monthly dividends totaled 44 cents per share for the year. The Fund's net asset value was $4.60 on Aug. 31, 1988, compared to $4.72 at the beginning of the fiscal year. If you purchased shares in the Fund during the year, your performance also would have been affected by the sales charge, as discussed in the prospectus.

Flexibility was an important part of the Fund's strategy, particularly after the October stock market drop. Immediately after the drop, we put our cash back into the market, in longer-term bonds to capitalize on falling interest rates. However, as inflation continued to rise throughout the year, we reduced our holdings in longer maturity bonds and moved toward securities with shorter maturities. The average maturity at the end of February was 22.6 years; by the end of August it was 15.2 years. Shorter maturity bonds are less affected by rising interest rates.

In another move to increase the Fund's defensive posture, we increased our exposure to mortgage-backed securities with short to intermediate maturities. These securities perform relatively well in a rising-interest-rate environment. At the same time, we have shifted the portfolio away from certain sectors—reducing our exposure in utility and high-grade industrial bonds with long maturities, as well as convertible bonds.

The IDS economic forecast calls for inflation to rise and interest rates on long-term Treasury bonds to peak sometime in the next year at perhaps 10.5 percent, up from 9.3 percent at the end of the fiscal year. This expected rise in rates requires us to continue to focus on more defensive sectors of the bond market, with a greater emphasis on short and intermediate maturities. By maintaining a defensive posture, we keep dividends as high as possible while minimizing the risks associated with rising interest rates.

Looking out over the longer term, we expect to see rates begin to decline in the second half of 1989 and we continue to maintain a positive outlook for the fixed-income markets.

FIGURE 35-6   CHANGE ORDER

CHANGE ORDER

Order No. _____ 1 _____
Date: _____ September 21, 1988 _____
Agreement Date: ____ August 1, 1988 ____

NAME OF PROJECT: ___ Contract 3–84 _____

___ Water Main Installation _____

OWNER: _____ City of Platteville _____

CONTRACTOR: _____ Lonsberg/Ochs Division of EAO Corporation ____

The following changes are hereby made to the CONTRACT DOCUMENTS:
   Add approximately 800 lin ft of 10-in water main for Main St. from Water St. to Broadway

Justification:
   Very old and deteriorated, must be replaced before street reconstruction

Change to CONTRACT PRICE:
Original CONTRACT PRICE $ _____ 53,860.00 _____
Current CONTRACT PRICE adjusted by previous CHANGE ORDER
$ ___ 53,860.00 ___
The CONTRACT PRICE due to this CHANGE ORDER will be
(increased) by: $ _____ 8125.00 _____
The new CONTRACT PRICE including this CHANGE ORDER will be
$ ___ 61,985.00 ___
Change to CONTRACT TIME:
The CONTRACT TIME will be (increased) by _____ 14 _____
calendar days.
The date for completion of all work will be ___ November 7, 1984 ___
(Date).

35.3

Approvals Required:
To be effective this Order must be approved by the federal agency if it changes the scope or objective of the PROJECT or as may otherwise be required by the SUPPLEMENTAL GENERAL CONDITIONS.

Requested by: _Merle Strouse_

Recommended by: Michael G. Lewis _Michael G. Lewis_

Ordered by: Merle Strouse

Accepted by: Edward Ochs _Edward Ochs_

● **EXERCISE 1** Prepare an inspection report on one of the following topics:

- The sprinkler system in the dormitory or other building
- Emergency evacuation procedures from the fieldhouse or auditorium
- The safety of ten popular toys
- The traffic flow patterns at a busy intersection
- The water purification procedures at your water treatment plant
- The crop harvest procedures of local farmers
- The construction status of a new commercial building
- The flame retardant properties of clothing
- The performance of a microwave oven
- The placement of lighting fixtures in classroom buildings
- The type of shopper (age, sex, and so forth) at a local store
- The college's access methods for handicapped persons

●**EXERCISE 2** Prepare an incident report on one of the following:

- A traffic jam during rush hour
- A home fire
- A bridge collapse
- The injury of a football, basketball, or other player
- An electrical power failure
- The flooding of a basement
- The removal of a large tree in a residential neighborhood
- The washout of a road culvert during a storm
- The collapse of a building because of snow or a storm
- The intentional demolition of a building
- The opening of a new store

●**EXERCISE 3** Obtain data from the local wastewater treatment plant and prepare a laboratory report on the efficiency of treatment.

●**EXERCISE 4** Monitor the progress at a construction site and prepare a report outlining the status of the project.

●**EXERCISE 5** Write a report detailing your progress in lowering your time for running the three-mile course around your campus. Include details of the overall exercise plan to run it in eighteen minutes or less and the progress to date.

●**EXERCISE 6** Write a report describing the traffic flow pattern at a major intersection. The number and types of vehicles per hour, the directions they come from, and turns made should all be included.

●**EXERCISE 7** Assume you are asked to write a periodic report of your progress toward completing your degree each semester as a condition of keeping your eligibility for athletics. Include information on courses completed, grades received, and cumulative grade point.

●36

FORMAL REPORTS

KEY POINTS

- Include a letter of transmittal with all formal reports.
- Develop a formal report that has an attractive appearance.
- Include a "Table of Contents," "List of Figures," and "List of Tables" at the beginning of a formal report.
- Write an introductory section, a procedures or methods section, and conclusions and recommendations for all formal reports.

Formal reports are the written communications sent to a client, funding agency, management, or another organization. They are often written to fulfill some contractual arrangement or are a part of the contract itself. Formal reports are frequently the outcome of months or years of data collection. As a result, the information contained within them is of more permanent value than that in an informal report.

The same attention to detail as accorded informal reports in Section 35.1, p. 223, should be given to formal reports. They should have an appealing appearance, contain accurate information, and be written in a clear and concise manner. However, the format of a formal report is often quite different from that of an informal one. Since the subject of the typical formal report is more complex, the format has more components.

36.1   INCLUDE ALL NECESSARY PARTS OF A TYPICAL FORMAL REPORT.

A formal report contains a front section, a body, a terminal section, and a back section. The typical format includes the following:

A. Front Section

1. Cover
2. Title page
3. Table of Contents
4. Lists of Figures and Tables
5. Abstract or Summary

B. Body

1. Introduction
2. Procedures
3. Results
4. Analysis
5. References

C. Terminal Section

1. Conclusions
2. Recommendations

D. Back Section

1. Appendixes
2. Index

An example of a formal report is provided at the end of this section. The sequence of steps for writing a formal report is different from the format sequence. These steps are as follows:

1. Analyze your audience.
2. Collect the data.
3. Write the body.
4. Write the terminal section.
5. Write the back section.
6. Write the front matter.
7. Write the letter of transmittal.

36.2   ANALYZE YOUR AUDIENCE.

Assess the needs and backgrounds of the intended readers of the report, then adjust the report's level of technical content accordingly. Often you can easily determine the readers' needs because they are outlined in the contract you have with them or with the organization requesting the study. The contract specifies what the organization wants to learn about a subject.

For example, you, an archeologist, are contacted by a city that is planning to extend a street through an area suspected of containing Indian artifacts. You are told the city intends to preserve sites containing significant cultural artifacts of earlier inhabitants. From this

245

information you know the purpose of the study, the needs of the reader, and the scope of the investigation.

Next, the city manager indicates that the intended readers of the study are the city council members who will take action on the proposed street extension. You then can determine what the background is of each of the city council members and gain an understanding of the level of technical language to use.

If instead you received a grant from the National Science Foundation to study the same site and were to report the results at a national meeting of archeologists, the technical level and terminology you use would be quite different. But you would have the same conclusions in each report about whether or not any significant cultural artifacts were at the site.

36.3 COLLECT THE DATA.

The second step in writing a formal report is to collect the data for the body of the report. You determine the purpose and scope of the study in your analysis of the needs and backgrounds of the readers. This helps you to determine your approach to the study and to know what information to collect.

36.4 WRITE THE BODY OF THE REPORT FIRST.

The body of the report includes an introductory section, procedures or methods section, results section, analysis section, and references.

Introductory section

The introductory section of your report consists of the purpose, scope, glossary and definitions, and organizational structure of the report.

Purpose   The first sentence(s) of the introductory section contains the specific objective(s) or purpose and subject of the report. Include sufficient background information for the reader to understand your report. Clearly and succinctly outline the reasons for your report and your understanding of the tasks that were to be completed.

Scope   The details and the limitations of your study are usually included in your scope. Since the study cannot provide an investigation of everything concerning its subject, clearly identify the limitations and report them in the scope. This will prevent the readers from misunderstanding the report. Although the scope should not be merely a listing of the items specified in the contract document, it is important that you include them here to indicate that the client is getting the contracted information.

The introductory section may repeat some information that the readers already know or have supplied to you, the writer. This is perfectly acceptable because the report should be self-sufficient. The report contains information that is of lasting importance, and someone who is unfamiliar with the project may refer to it in the future.

Glossary and Definitions   If necessary, include a glossary and definitions for the convenience of the readers. A glossary may contain special terms, abbreviations, or symbols in the report, with extensive technical, mathematical, or scientific analysis. It may be inserted either as part of the back matter or as part of the front matter. Do not put the information in both places.

However, provide a list of the terms you use and their definitions in the introductory section. Often a technical report has terms that are unfamiliar to readers but are necessary for a complete understanding of the report. Each technical field may have unique abbreviations or idioms that are customarily used. These may make the report easier to read than if you used a whole phrase each time it is repeated or if you had to explain each idiom. The abbreviations and idioms may be unfamiliar to the reader, so they should be explained in a list near the beginning of the report.

Procedures or methods section

Adjust the details of the procedures followed or methods used to fit the needs of the report and its readers. How the report's information was collected is often explained in its introductory portion. In other reports the procedures may be described in detail in their own section. The methods in a laboratory report or pilot plant study, for example, may require more detail than other types of reports. This is particularly true if knowledge of the procedures is important. Otherwise, the methods used may be only referenced to a book or manual of standard procedures. See Figure 35-4, p. 236, where only the reference, *Standard Methods*, is given.

In many cases a report is written for nontechnical readers who are not as concerned with the details of procedures as with the general approach to the problem. If so, modify the details to fit the situation.

Describe in the procedures section the equipment used in the study. Provide only the general operating procedure. For a technical audience, place more detailed procedures in an appendix for reference.

Discuss in this section the underlying theory or physical law associated with the subject of the report. Describe, for example, the derivation of equations necessary for the study or discuss those taken from other sources in sufficient detail for the reader to understand the results.

Results section

Organize the results in a logical manner so that the reader can understand them. Your results section should contain laboratory data, results of field investigations, and other information you have gathered. If the information consists of a large amount of numerical data, you will find it is often more understandable presented in a table or graph. You should, however, still describe the information in the text that accompanies the tables or graphs. Do not place the burden on your readers to pick out the important information; tell them what the important points are. Because the data's relationships and trends are often more apparent in an illustration than in the text alone, the descriptive text can be reduced. Section 39 has more information on the organization of tables and graphs.

Include in the report only information that leads directly to the solution of the problem. Leave out of the main body of the report information unrelated to the major purposes of the study. If the peripheral information may be of interest to readers, you can always include it in an appendix. Some projects may yield detailed test data that are too extensive to be included in the body of the report itself. A summary table and discussion would be appropriate for the results section, but the details should be included in an appendix.

Analysis section

The analysis section is one of the most important parts of your report. This is where you present the answer(s) to the problem statement(s). The discussion or analysis section follows directly after the results. Include a careful and logical discussion for each portion of the results. Develop each point in one or more paragraphs, discussing no more than one point in a paragraph.

Often the results and a discussion of them are contained in the same section. In that case, after each point is made, discuss the data to support it. Remember that when you combine these sections, you present and discuss each graph or table containing the data before you introduce the next table or graph.

Data relationships and trends are more apparent in a graph than when presented in a table or in the text alone. To illustrate this point, the data from the report at the end of this section are included in Table 36-1 and in Figure 36-1. These data are from a pilot plant study of two filters. The removal of BOD (Biochemical Oxygen Demand) and TSS (Total Suspended Solids) by the filters is presented in the graphs. It can be seen that filter B is better and consistently removes more BOD and TSS than does filter A. The effluent concentration is lower from filter B than from filter A. When presented in Table 36-1, the relationship is not as clear as in the graphs. The graphs show the relationship and trend in filter removal over time and make it easy to describe this in the discussion section that follows.

The analysis in the report at the end of this chapter goes beyond the pilot plant results. Filter Company A and Filter Company B submitted a design for a full-scale filter. The costs of each of the two filters are determined and discussed as a part of the analysis in a separate section because the cost comparison is different than the performance comparison. Since only the costs are compared, a table is the appropriate graphic to use at the end of the section on costs. (See Table 5.3-1 in the sample report.)

References

The references are the sources from which you derived information to complete the report. They are placed in this section so that they follow immediately after their citations. The format for documentation is covered in more detail in Section 40.

TABLE 36-1 DATA DISPLAYED AS A TABLE

RAPID SAND FILTER PILOT TEST DATA
WHEATON SANITARY DISTRICT

DATE	BOD MG/1		
		Effluent	
	Filter* Influent	Co. A	Co. B
4/27–28/81	118.	—	—
4/28–29/81	66.	—	22.
4/29–30/81	50.	—	12.
4/30–5/1/81	45.	30.	18.
5/1–2/81	48.	28.	25.
5/2–3/81	51.	—	24.
5/3–4/81	51.	—	24.
5/4–5/81	68.	—	—
5/5–6/81	60.	46.	—
5/6–7/81	46.	—	38.
5/7–8/81	66.	—	45.
5/8–9/81	80.	64.	46.
5/9–10/81	84.	61.	49.
5/10–11/81	81.	40.	34.
5/11–12/81	60.	40.	26.
5/12–13/81	57.	47.	27.
5/13–14/81	68.	54.	34.
5/14–15/81	—	—	—
5/15–16/81	46.	35.	25.
5/16–17/81	46.	38.	28.
5/17–18/81	63.	55.	39.
5/18–19/81	100.	39.	37.
5/19–20/81	96.	50.	41.
5/20–21/81	92.	56.	42.
5/21–22/81	130.	66.	43.
5/22–23/81	70.	54.	38.
5/23–24/81	96.	54.	48.
5/24–25/81	86.	63.	41.
5/25–26/81	68.	66.	51.
5/26–27/81	86.	68.	51.
5/27–28/81	122.	74.	57.
5/28–29/81	84.	84.	52.

*Primary Influent

| | TSS MG/1 | | HYDRAULIC LOADING GPM/SQ FT | |
| | Effluent | | | |
Filter* Influent	Co. A	Co. B	Co. A	Co. B
86.	32.	23.	.1.1	2.3
130.	41.	24.	2.1	2.3
72.	—	4.5	—	2.3
38.	21.	7.3	1.9	2.3
38.	15.	4.	1.9	2.3
32.	11.	4.	2.0	2.3
38.	—	9.3	—	2.3
60.	—	—	—	—
52.	18.	13.	1.5	2.3
70.	24.	18.	2.1	2.3
52.	20.	12.	2.0	2.3
54.	29.	11.	2.0	2.3
50.	38.	11.	2.0	2.3
86.	24.	16.	1.9	2.3
142.	17.	8.3	1.9	2.3
50.	20.	10.	2.0	2.3
70.	22.	12.	2.0	2.3
90.	20.	11.	1.0	2.3
56.	17.	14.	1.6	2.3
34.	21.	9.3	2.0	2.3
54.	26.	14.	2.1	2.3
100.	15.	14.	2.1	2.3
88.	29.	18.	2.0	2.3
88.	28.	14.	2.0	2.3
102.	43.	16.	2.1	2.3
96.	22.	17.	2.0	2.3
38.	18.	15.	2.0	2.3
96.	37.	20.	2.0	2.3
118.	27.	19.	1.9	2.3
52.	35.	18.	2.0	2.3
90.	26.	19.	2.0	2.3
62.	35.	18.	2.0	2.3

*Primary Influent

36.5

FIGURE 36-1 PILOT PLANT DATA DISPLAYED IN A GRAPH

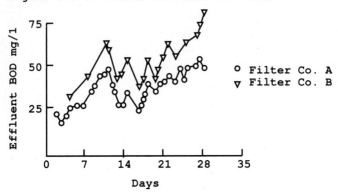

Figure 3.1-1 Filter BOD Data vs. Time

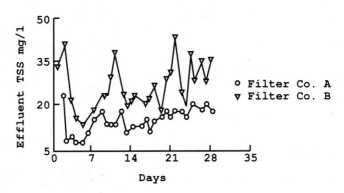

Figure 3.1-2 Filter TSS Data vs. Time

36.5 PLACE THE CONCLUSIONS AND RECOM-MENDATIONS IN A SEPARATE, OFTEN TERMINAL, SECTION OF A REPORT.

Conclusions are often a concise summary of report findings. You already may have mentioned them in the analysis section, but summarize the main points here for added emphasis. Be specific in brief paragraphs; do not introduce new material or further analysis in this section. Do not repeat tables or illustrations here that are included in the analysis or results section.

The recommendations are actions that should be taken as a result of the study. Since they follow logically from the conclusions, you may include them in the same section or you may set them apart in their own section.

The location of the conclusions and recommendations section within a report varies depending on your preference and the anticipated needs of the intended readers. The several locations where conclusions might be placed are:

1. as part of the front matter,
2. ahead of the introduction as the first section,
3. as the next section after the introduction, or
4. as the last section of the body.

If your report is intended for a management audience, include the conclusions and recommendations near the beginning. Management is interested in the alternatives and the advantages and disadvantages of each. Placing the conclusions and recommendations near the beginning makes it easier for readers to find the main points. They can then decide if more investigation of the supporting data is necessary and are better able to follow the arguments of the discussion section.

The position of the conclusions governs what you include in other parts of your report. If you place them near the beginning, you do not need a separate summary at the end, nor do you need to include detailed results in the abstract. If they are placed after the analysis section, you can include a brief summary statement in your letter of transmittal or in your introductory section.

In the report on filters, the recommendation is included after the discussion of the results. A brief summary of the conclusions is used to explain the reason for the recommendation to select Filter Company B. This recommendation develops logically from the conclusions reached for each of the bases of comparison: cost and performance.

36.6 INCLUDE ALL NEEDED INFORMATION IN THE FRONT SECTION.

The front matter in a technical report precedes the body. The purpose of the front section is to help the reader find information contained within the report. The front section may include some or all of the following: cover, title page, table of contents, lists of figures and tables, and abstract or summary.

The cover

The purpose of the cover is twofold: to protect the report and to aid readers in identifying the report once it is on the shelf. The cover is usually made of a heavier stock paper than the rest of the report. Consulting firms often use colored stock for the cover and use the same color for all reports of the same type.

The minimum information that you should include on the cover is the report title and the name of the originating agency. Companies may also include their letterhead or other company symbol. Reports prepared for government agencies may also include security markings and the report series number. Your cover may have the identifying information printed directly on it or have a window cut out so that the same information contained on the title page is visible through the cover. Figure 36-2 and Figure 36-3 are examples of report covers from a consulting university and from a government agency.

The title page

The title page is the first page inside the cover of a report. The title is exactly the same as on the report cover. The title page should have a pleasant appearance with appropriate spacing of the information. The first impression of your report is important because the reader makes a judgement about the report content based on the physical appearance and content of the first two pages. In addition to the title, the title page should contain your name and affiliation, the client's name and location, and the publication date. Include the contract number, approval signatures, engineering seal, and security markings, if appropriate. See Figures 36-4 and 36-5 for examples of title pages.

FIGURE 36-2 REPORT COVER

A DIRECTORY OF COMPUTER PROGRAMS
APPLICABLE TO U.S. MINING PRACTICES
AND PROBLEMS

Prepared for
UNITED STATES DEPARTMENT OF THE INTERIOR
BUREAU OF MINES

by
THE UNIVERSITY OF WISCONSIN-PLATTEVILLE
DEPARTMENT OF MINING ENGINEERING
PLATTEVILLE, WISCONSIN 53818

FINAL REPORT
on
Contract No. G0264026
A Directory of Computer Programs Applicable to U.S. Mining
Practices and Problems

November 1977

FIGURE 36-3 REPORT COVER

United States
Environmental Protection
Agency

Technology Transfer   EPA/625/4-85/016

 Seminar Publication

Protection of Public
Water Supplies from
Ground-Water   Contamination

FIGURE 36-4   TITLE PAGE

Technology Transfer   EPA 625/2-79-024

Capsule Report

Acoustic Monitoring To Determine the Integrity of Hazardous Waste Dams

August 1979

**This report was developed by the
Industrial Environmental Research Laboratory
Cincinnati, OH 45268**

FIGURE 36-5   TITLE PAGE

EFFECTS OF COAL DEVELOPMENT IN
THE NORTHERN GREAT PLAINS

A Review of Major Issues and Consequences
at Different Rates of Development

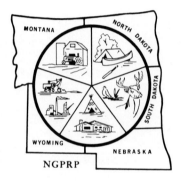

Prepared as Part of the NGPRP (Northern Great
Plains Resources Program) in Cooperation
with Federal, State, Regional, Local,
and Private Organizations

April 1975

Your title should provide as much information as possible about the subject of your report while still being brief. It should be clear and to the point. Your reader should not be misled because the title is too short, or miss the point because the title is too long and wordy.

WEAK TITLES	IMPROVED TITLES
Sludge Disposal	Environmental Assessment of Sub-surface Disposal of Municipal Wastewater Treatment Sludge
The Prevalence of Sub-surface Migration of Arsenic, Selenium, PCBs, Aldicarb, and Lead at Selected Municipal, Industrial, and Abandoned Waste Land Disposal Sites	Migration of Hazardous Chemical Substances at Land Disposal Sites

Capitalize the first word in the title and all other words except articles, coordinating conjunctions, and short prepositions (see Section 19). If the title is longer than one line, it should be double- or triple-spaced.

The table of contents

The table of contents follows the title page and contains a topical outline of the report. It provides a guide for the reader to quickly locate parts of the report and the pages on which they occur. The table of contents also allows readers to determine the relative weightiness of topics. The section headings that appear in the report are exactly the same as the corresponding items in the table of contents.

Since the table of contents is an outline of your report, include appropriate subheadings. Indent them to indicate their rank compared to the major headings. Usually, no more than three levels of subdivision of the main topics are necessary in the table of contents. You may use more subdivisions in the body of the report itself, but don't clutter the table of contents.

Begin placing the front material page numbers on the table of contents page. Indicate them with small Roman numerals (i, ii, iii, iv, v, and so forth). Do not number the letter of transmittal when it is inside the report or the title page, but they are counted. The table of contents is never on page *i*.

Underline and center the words "Table of Contents" at the top of the page. Leave three or four spaces below this before listing your first entry. Doublespace between the major headings. Single or doublespace between subheadings. The point is to make the table of contents have a pleasant appearance and provide the reader with helpful information.

Authors use several different systems of outlining to organize the table of contents. No one system is better than another as long as it meets the needs of both the author and the readers. The traditional approach is to use capital Roman numerals for the major headings, capital letters for the first level of subdivision, and Arabic numbers for the second level of subdivision. Figure 36-6 is an example of this type of table of contents.

Some authors use only the Roman numerals or capital letters for the major headings and rely on indenting the second-order headings five spaces and the third-order headings ten spaces from the left margin to give structure to the outline. Figure 36-7 is an example of this type of table of contents.

A third method of organizing the table of contents in a report is to use the Arabic decimal system. The first topic is Section 1, the subtopics are labelled Section 1.1, 1.2, 1.3, and so on; the third level of subdivision is 1.1.1, 1.1.2, 1.2.3, and so on. Figure 36-8 is an example of this type of table of contents.

Place the page numbers in a table of contents opposite each heading in a column on the right-hand side of the page. Prepare the Table of Contents page after the report is in final form so that it has exactly the same headings as the report and the correct page numbers. Place a series of dots between the last word of a heading and the page numbers at the right-hand margin of the page. Either single- or double-space the dots. They are to help readers easily find a page number for a section. Accordingly, the last word in a title should not extend all the way to the page numbers. Leave adequate space between the end of a title and the corresponding page number to prevent confusion.

List of tables and list of figures

Include a "List of Tables" and "List of Figures" if the report contains more than six tables or figures. List the tables and figures in the order in which they occur in the report. Exactly the same table and figure numbers and titles should appear in the lists as in the text of your report. Be sure the titles of the tables and figures are concise and informative.

The figure or table number corresponds to the chapter or section number. Capitalize all important words in the titles, just as in the table of contents. Single-space any title that continues on more than one line. Do not stretch the titles all the way to the right-hand margin. Make sure the table and figure page numbers can be easily distinguished from the title by leaving adequate space between the two. Place the page numbers in a vertical column along the right margin of the page, across from the end of each title.

FIGURE 36-6 TABLE OF CONTENTS EMPLOYING A TRADITIONAL OUTLINE FORMAT

TABLE OF CONTENTS

FIGURE 36-7 **TABLE OF CONTENTS EMPLOYING CAPITAL LETTERS FOR HEADINGS**

TABLE OF CONTENTS

FIGURE 36-8 **TABLE OF CONTENTS EMPLOYING ARABIC DECIMAL SYSTEM**

TABLE OF CONTENTS

    As in the table of contents, insert single- or double-spaced dots between the end of the title and the page number. Titles should be double-spaced to clearly separate them. All lists may be placed on the same page if space permits, and if the report is short, you may include the lists on the table-of-contents page. A List of Figures and List of Tables are included in the sample formal report at the end of this section.

Abstract or Summary

As indicated in Chapter 34, an abstract is like an outline of the report. It keeps the same order as the report and covers the significant points. A summary, on the other hand, does not necessarily follow the same order as the report and only emphasizes the main points being made.

Include a summary in an appropriate location in the report. A summary is often part of the front matter of a report to provide the reader with easy access to the report's main points. More readers look at the summary than any other single part of the report, often using the information to decide if reading the whole report is worthwhile. Management people, who are usually too busy to read the whole report and are only interested in the main points, read the summary. Others who have more time will use the summary to focus on the key points before they read the other parts of the report.

Include in the summary enough introductory material for the reader to learn what your report is about, how your study was conducted, and what your conclusions are. You will find that a summary is more easily written after the main body of the report is completed. Include only information contained in the report; introduce no new information. If the report will be read by both a technical and nontechnical audience, write the summary for nontechnical readers. Summaries are usually no longer than two pages unless a report is extremely long (greater than one hundred pages). See Section 34 for a more complete discussion of the summary.

36.7 INCLUDE AN APPROPRIATE BACK SECTION.

The back section consists of material that is unnecessary for a complete understanding of the report but may be desirable as supporting information. The back matter usually is divided into two sections: appendixes and index.

Appendixes

The appendixes provide information that is too extensive or detailed to include in the main body of the report. Appendixes may contain the original data obtained during the study; because of a large amount of data, only a summary table would be included in the body.

Mathematical derivations and sample calculations are sometimes included in an appendix, as are data collection sheets and standard test forms. Also insert documentary papers, such as the transcript of a public hearing, letters, affidavits, or other legal papers, in an appendix. Provide a separate *Appendix* for each type of information. The table of contents should have a separate title and page number for each appendix. The appendixes are often labelled *A*, *B*, *C*; and the pages numbered *A1*, *A2*, *A3*, *B1*, *B2*, *B3*, and so forth to keep them separate from the rest of the report. They may be further set off by a separate page titled *Appendixes*.

The sample report on sand filtration contained three appendixes. These were omitted in this textbook to save space.

Index

An index is an alphabetical listing of all significant topics, subjects, or phrases mentioned in a report. Provide the page number for each entry. Only large reports with a wide range of topics, wide and frequent usage, and hundreds of pages need an index.

36.8   ACCOMPANY ALL FORMAL REPORTS WITH A LETTER OF TRANSMITTAL.

The purpose of a letter of transmittal is to introduce a report. Your letter should follow the format of business letters discussed in Section 31. The main elements of this conversational business letter are the letterhead, inside address, salutation, body, and closing. The letterhead contains the name of your firm, organization, or agency responsible for the report, and the date. The inside address should have the name, title, and address of the recipient. The formality of the salutation depends on the relationship between you and the reader. The body of your letter should include the purpose of your report, a brief outline of the problem, a brief summary of the findings, and some concluding remarks. Your closing should be formal with your signature, official title, and organization.

Write the letter of transmittal at a technical level appropriate for your reader. Be sure it is grammatically correct with no spelling or typing errors. This is usually the first item that the reader sees and the first impression you will make about the competence of both you and the organization you represent. The accuracy of your report and competence of your organization may be questioned if the cover letter has any mistakes.

The letter can be either attached to the outside of your report or bound in the report. If it is bound in the report, place it inside the cover or just after the title page. Bind it so that it doesn't become separated from the report. In this way it can serve as an introduction to your report for future readers who are not familiar with the context of the report preparation.

An example of a typical letter of transmittal is provided in the sample formal report at the end of this section.

●**EXERCISE 1.** Prepare a formal report on one of the following topics:

- The effects of caffeine on the human body
- The feasibility of recovering energy from incinerated solid waste
- Evaluate and determine the best videocassette recorder
- The impact on the student athlete of raising academic standards
- The adequacy of catastrophic illness insurance
- Effective measures for controlling employee theft
- The adequacy of smoke detection systems in a local college dormitory system
- Evaluation of the methods for controlling water seepage in basements
- The cost effectiveness of street cleaning
- The feasibility of establishing or expanding a research center in your field at your college
- The adequacy of energy conservation measures at your college
- The adequacy of soil erosion control programs
- The investment alternatives available for someone with $25,000
- The cost effectiveness of construction of a new bridge
- The adequacy of the storm warning system in your area

August 25, 1981

Wheaton Sanitary District
15640 Shaffner Road
Wheaton, Illinois 60187

Letter of
transmittal

Attention: Mr. Robert L. Clavel, P. E.
             Engineer and Manager

Re: Rapid Sand Filter Pilot Study
      Evaluation of Study Data and Proposals

Dear Mr. Clavel,

In accordance with our agreement dated
October 31, 1980, we have completed an
evaluation of pilot test data and proposals
submitted by Filter Company A and Filter
Company B for providing rapid sand filter
equipment as part of the District's proposed
facility upgrading. Based on our analysis of test
results and system costs, as outlined in the
attached report, we recommend award of the
Filter Equipment Contract to Filter Company B.

Reminder of
what study was
about and recom-
mendations

We would like to thank the District staff, whose
efforts and support aided in completing this work.

Sincerely,
STRAND ASSOCIATES, INC.

Michael D. Doran

Michael D. Doran, P.E.

Roger H. Huchthausen

Roger H. Huchthausen, P.E.

MDD/RHH:ME

Title page of the
report

WHEATON SANITARY DISTRICT

REPORT ON SAND FILTER PILOT STUDY

STRAND ASSOCIATES, INC.
Consulting Engineers
910 West Wingra Drive
Madison, Wisconsin 53715

August 1981

TABLE OF CONTENTS

LIST OF TABLES

LIST OF FIGURES

SECTION 1

INTRODUCTION

1.1 PURPOSE AND SCOPE OF STUDY

Introductory
section

Purpose and
scope

Background of
study and need
for the study

The purpose of this study and the subject of this
report is the evaluation of two rapid sand filters to
be used for filtration of untreated municipal
wastewater. The proposed improvements to the
Wheaton Sanitary District wastewater treatment
facilities include the construction of rapid sand
filters. The sand filters would normally be used to
filter clarified effluent from the existing
nitrification facilities. During periods of high wet-
weather flow, however, the filters would be used to
filter primary effluent prior to disinfection and
discharge along with clarified and disinfected
effluent from the nitrification facilities. During
normal flow periods, the filters would be required
to meet maximum monthly average
concentrations of 10 mg/l BOD and 12 mg/l TSS,
while filtering clarified effluent from the
nitrification facilities. During high wet-weather
flows, while filtering primary effluent, the filters
would be required to meet monthly average
maximum concentrations of 30 mg/l BOD and
30 mg/l TSS.

Because of the requirement to filter primary effluent during high wet-weather flows, and since primary effluent filtration is a relatively new concept, the District elected to require prospective equipment manufacturers to prequalify their equipment by on-site performance testing, and to receive bids from those manufacturers successfully completing on-site tests.

A pilot plant is a smaller-scale working model.

Pursuant to the District's "Notice to Prospective Bidders" and "Qualifications Information Package," dated February 11, 1981, the District received notification from two manufacturers of their intent to conduct on-site pilot testing and to submit a bid for providing necessary sand filter equipment as part of the treatment facility expansion. These manufacturers were Filter Company A and Filter Company B. Both manufacturers subsequently conducted concurrent pilot testing and submitted bids for providing specified equipment.

1.2 ABBREVIATIONS

The following abbreviations have been used:

Terms typically abbreviated in this type of study

BOD	five-day biochemical oxygen demand
gpd	gallons per day
gpm	gallons per minute
mgd	million gallons per day
mg/l	milligrams per liter
NPDES	National Pollutant Discharge Elimination System
O&M	Operation and Maintenance
sq ft	square feet
TSS	total suspended solids

SECTION 2

TEST EQUIPMENT AND PROCEDURES

Pilot-scale filtration equipment and operating personnel were provided by each manufacturer during the period of the test. The units were tested concurrently with a common source of wastewater on a "side by side" basis. The filtration units were tested for four weeks with primary effluent.

2.1 PILOT TEST EQUIPMENT

The Filter Company B unit consisted of a single filter cell of 4 sq ft and included clearwell, mudwell, influent and backwash pumps, air blower, automatic effluent sampler, and automatic controls. The filter cell was operated with 11 inches of 0.5 mm effective size sand.

The Filter Company A unit provided 34 sq ft of total surface area with seventeen 2-sq-ft cells. The media used was 11 inches of 0.6 mm effective size sand.

Influent to the pilot filters was provided by a 200-gallon head box. The head box received primary effluent or unchlorinated final effluent as required during the tests. Primary effluent was delivered to the head box by means of a submersible pump placed in the plant final-effluent-diversion structure. In this manner, both pilot filters were provided with a common influent source.

2.2 SAMPLING AND ANALYTICAL PROCEDURES

Standard
procedures
reference

Influent and effluent samples were collected by means of automatic sequential samplers, which collected a time-composite of hourly samples throughout each sampling day. Samples were kept on ice during collection, and all analyses were performed in accordance with the 14th Edition of *Standard Methods* (American Public Health Association, 1975). All samples were analyzed by the Wheaton Sanitary District staff as soon as possible after collection.

SECTION 3

PILOT TEST RESULTS

Results section

3.1 TEST DATA

Figures 3.1-1 and 3.1-2 show BOD and TSS data vs. time for the primary effluent filtration tests.

3.2 DISCUSSION OF RESULTS

Discussion of
results

As indicated in Figures 3.1-1 and 3.1-2, the Filter Company B filter performed significantly better than the Filter Company A unit during the primary effluent filtration tests. It is anticipated that primary effluent BOD and TSS will fall within the range of 40–80 mg/l during wet-weather flow periods. For the test data within this range, the median of the Filter Company B filter effluent data was 27 mg/l BOD and 12 mg/l TSS, whereas the median of the Filter Company A filter effluent data was 46 mg/l BOD and 24 mg/l TSS. The required filtered primary-effluent maximum monthly average BOD and TSS concentrations in the District's NPDES discharge permit are both 30 mg/l. Therefore, it is concluded that the Filter Company B unit met the primary effluent filtration requirements, whereas the Filter Company A unit did not.

SECTION 4

PROPOSED DESIGN CRITERIA

4.1 FILTER COMPANY B

Based on the pilot test results, Filter Company B recommended provision of six filter cells, each of 372 sq ft, for a total surface area of 2,232 sq ft. The units would be shallow bed with a single-graded fine sand media, and would be provided with surface-air-mix, chemical cleaning systems, and pulse-bed capability. Recommended design criteria are listed in Table 4.1-1.

Proposal by companies based on results

4.2 FILTER COMPANY A

Filter Company A has recommended provision of four filter units each of 896 sq ft, for a total filter surface area of 3,584 sq ft. The units would be shallow bed with a single-graded fine sand media. Each filter would be constructed to provide 84 individual filter cells of 10.67 sq ft each. Recommended design criteria are listed in Table 4.2-1.

TABLE 4.1-1 FILTER COMPANY B DESIGN

Number of Units	6
Size of Units	12 ft wide × 31 ft long
Total Surface Area	2,232 sq ft
Hydraulic Loading Rate	1.77 gpm/sq ft
Depth of Media	10 in
Size of Media	0.45 mm, Effective Size Uniformity Coefficient 1.5
Backwash Rate	12 gpm/sq ft
Air Mix Rate	58 cfm/1000 cu ft at Media

TABLE 4.2-1 FILTER COMPANY A DESIGN

Number of Units	4
Size of Units	16 ft wide × 56 ft long
Total Surface Area	3,584 sq ft
Hydraulic Loading Rate	1.10 gpm/sq ft
Depth of Media	11 in
Size of Media	0.55 mm to 0.65 mm, Effective Size Uniformity Coefficient 1.5
Backwash Rate	15–30 gpm/sq ft

SECTION 5

ESTIMATED COSTS

5.1 CAPITAL COSTS

Estimated capital costs for both systems are shown in Table 5.1-1. Costs are based on equipment manufacturer's bid prices for equipment and estimated equipment installation costs. Estimated costs for structures including process piping, plumbing, disinfection, and post-aeration were not included as these were considered common to both systems.

Analysis of
results and
proposals

5.2 OPERATION AND MAINTENANCE COSTS

Operation and maintenance costs were compared based on estimated differences in power and chemical costs for the two systems. Estimated operation and maintenance cost differences are shown in Table 5.2-1. Costs for labor and equipment maintenance are not included, as these costs are considered to be approximately equal for both systems. Costs are based on treating an average of 7.5 mgd during the design period. As indicated in Table 5.2-1, operation and maintenance costs are estimated to be somewhat higher for Filter A.

TABLE 5.1-1 ESTIMATED CAPITAL COSTS

ITEM	FILTER B	FILTER A
Equipment	$696,000	$618,000
Structures	779,000	681,000
Heating and Ventilating	40,000	35,000
Electrical	60,000	55,000
SUBTOTAL	$1,575,000	$1,389,000
Technical Services and Contingencies @ 25%	394,000	347,000
TOTAL CAPITAL COST	$1,969,000	$1,736,000

Note: All costs in 3rd-quarter 1981 dollars. Third-stage pumping, disinfection, and post-aeration not included.

TABLE 5.2-1 ESTIMATED OPERATION AND MAINTENANCE COSTS

ITEM	FILTER B	FILTER A
Power	$7,000	$5,000
Chemicals	11,000	28,000
Treating Filter Backwash Water	31,000	37,000
TOTAL	$49,000	$70,000

5.3 TOTAL PRESENT-WORTH COSTS

Total present-worth costs of the two systems are compared in Table 5.3-1. A discount rate of 7.125 percent for twenty years was used. A salvage value of 50 percent of initial capital costs for structures was used (present-worth factor = 0.252). As indicated, the two systems are estimated to be essentially equal in present worth, with higher capital costs for the Filter Company B system being offset by lower operating costs.

TABLE 5.3-1 COMPARISON OF TOTAL PRESENT-WORTH COSTS

	TOTAL PRESENT-WORTH COST	
ITEM	Filter Co. B	Filter Co. A
Equipment	$870,000	$772,000
Structures	974,000	851,000
Heating and Ventilating	50,000	44,000
Electrical	75,000	69,000
Subtotal	$1,969,000	$1,736,000
Salvage Value	−123,000	−107,000
Increased O & M Costs	—	220,000
TOTAL PRESENT WORTH	$1,846,000	$1,849,000

Note: All costs in 3rd-quarter 1981 dollars. 7.125% discount rate for 20 years used. Capital costs for third-stage pumping, disinfection, and post-aeration not included. Labor and maintenance costs common to both systems not included.

SECTION 6

CONCLUSIONS AND RECOMMENDATIONS

Terminal section   As discussed in Section 3, the Filter Company B system performed substantially better than that of the Filter Company A system, and met all effluent requirements. The Filter Company A unit, however, failed to meet treatment requirements during primary effluent filtration.

Based on demonstrated successful performance and essentially equivalent total present worth costs, as discussed in Section 5, award of the Filter Equipment Contract to Filter Company B is recommended.

●37

PROPOSALS

KEY POINTS

- Be sure to use a format acceptable to the person or group for whom the proposal is written.
- Always strive for neatness and accuracy in a proposal.
- Include all needed information in the introduction, body, and summary of the proposal.

A proposal is a written request to sell a service or a product or to receive funding to work on or complete a project, such as lab research. Proposals may be in-house (internal within the company) documents or documents written for a customer, but they usually have one common denominator: money.

Students also write proposals—to obtain funding for research but more often to seek acceptance for a course of study. Obviously, money is rarely involved with this type of proposal.

Short proposals are often presented as memos (Section 32) or letters (Section 31), while long proposals are usually formal reports (Section 36) that may run hundreds of pages. Whatever the length or format, certain characteristics are true of most proposals, for they should include very specific information.

37.1   INCLUDE ALL NEEDED INFORMATION IN THE INTRODUCTION, BODY, AND CONCLUSION.

a. State the subject, purpose, statement of the problem, background, need, and scope in the introduction.

Subject and purpose   In the subject you state *what* you will be examining while in the purpose you explain *why* you plan to examine it. The subject and purpose are sometimes combined as the *objective.* As a general rule, use separate subject and purpose statements when your proposal is long or involves a highly technical subject, and use a combined statement when your proposal is short or deals with a simple problem.

Statement of the problem   Often a proposal states exactly what the problem is. If you want to, you can place this in a special section entitled *Statement of Problem.* Define and describe the problem clearly in this section.

Background   The background section gives the sequence of events that has led to the current problem. Consider what has happened historically. Give a brief account that highlights the most important events, especially ones in the recent past. By doing this, you demonstrate your grasp of the situation as well as provide enlightening material for the reader.

Need   Your proposal must show that a strong need for a change or for a product exists. You have to convince your reader that your study or product is necessary.

Scope   You must state what is going to be covered in your proposal. You also may want to include what you will not do as well as what you will do, for the scope protects both you, the proposal writer, and your client against any misunderstanding about what to expect.

b. State the methods, timetable, qualifications, cost, feasibility, and expected results in the body of the proposal.

Methods   Explain in the proposal how your stated plan will be carried out. You must answer the question *how* and show that you have the necessary means to complete your work.

Note: *Methods* are sometimes replaced by a *Procedures* section. Occasionally a long proposal will include both. A procedures section is usually a methods section expanded in greater detail.

Timetable   Most proposals include a timetable for accomplishing whatever is being proposed. Your timetable may be a limited one or it may extend for a year or more. Sometimes you will need a strict timetable with dates of completion; at other times you will only need a deadline date. Always be sure your deadlines are realistic ones.

Qualifications   Many proposals require that the qualifications of the personnel at a company or the qualifications of the researcher be included either in the proposal or in resumes attached at the end of the proposal. Include previous experience and, if required, list personnel references in this section.

Cost   The main reasons for writing most proposals is to obtain funding or payment. Occasionally your costs are included on a separate sheet; usually they are included in the proposal. Whichever the case, be sure to make the costs clear and binding.

Feasibility   Sometimes an idea sounds feasible but in reality it is not feasible, perhaps because it is too expensive or too unusual. Your proposal should demonstrate its feasibility, if needed. Considerations to include are whether your plans will solve a problem, are practical, and are desirable.

Expected results   Your proposal needs to explain what will be obtained from the study. You must commit yourself to what you are going to deliver.

c. Conclude with a summary, if the proposal is long, and a call to act on the proposal.

Summary   Formal proposals begin with an abstract or summary (see Section 34 for a complete discussion); however, many professional proposals need to end with a clear, well-developed summary to reiterate the main points.

Call to action   Your proposal needs to end in a positive, persuasive way to encourage immediate action on it. Because of their length, long proposals often begin with a "call to action"; sometimes writers even choose to place this in the abstract or summary to make certain that everybody reads this part and that it does not get buried at the end. In a short proposal, place the call to action at the end to persuade your reader *after* you have presented your facts.

37.2 SELECT THE APPROPRIATE FORMAT FOR THE PROPOSAL.

Proposals exist in various forms. Determine the correct format for a proposal by one of the following means:

1. The agency requesting the proposal specifies a format you must follow, which is called a Request for a Proposal (RFP),
2. the length of the proposal determines the format, or
3. you base your decision on what you know about the group receiving your proposal.

In the student proposal in Figure 37-1 at the end of this section, the student followed a format given to her by her teacher. In the faculty proposal in Figure 37-2, Dr. Roger A. Nelson was bound by a specific format published by the agency who requested the proposal.

37.3 MAKE SURE THE PROPOSAL IS NEAT, WELL-WRITTEN, AND ACCURATE.

a. Neatness

The appearance of your proposal is extremely important. It may have many selling points, but if it is not displayed in an attractive format, you will not impress the person or group to whom it is directed. Be sure to submit your proposal on high-quality bond paper. Neatly and correctly type it or have it printed. If it is longer than four or five pages, you should bind it with a fairly heavy paper cover. Some companies use loose-leaf binders while others have their proposals spiral-bound.

b. Style

Your proposal should be clear and free from spelling and typographical errors. Employ correct grammar, usage, and punctuation, and conventional mechanics.

37.3

Of equal importance are your word choices. Write your proposal in a positive manner, tailored for your audience (Section 21 has a complete discussion of audience types). Above all, show enthusiasm for your subject.

In the sample proposals, both the student and Dr. Nelson write with enthusiasm. Although they use fairly formal tones, their interest and commitment to their projects are evidenced throughout their proposals. In addition, both writers are clear and refrain from using confusing technical terms. When Dr. Nelson must use a technical term, he does so in a clear context, or he defines the term.

c. Accuracy

Many proposals employ exact figures or specific equations. These must be accurate, as must be the research (library, previous studies, field work, and the like) presented in the proposal. Any inaccuracies only point to sloppy research or careless proofreading. And these are not what a person or group asking for a proposal is seeking!

●**EXERCISE 1** The two sample proposals that follow these exercises are very different from one another. In examining them, notice that the audiences and the purposes for writing the proposals play important roles in determining the way they are both written and formatted. Explain the basic differences in the organization of the two proposals and the reason these differences occur.

●**EXERCISE 2** Obtain a copy of a proposal that has been accepted for funding. A professor or a business acquaintance may be able to furnish one. If not, find one in the school library. Bring a copy of the proposal to class to compare it with the ones other class members bring. In what ways do the proposals differ in purpose? In organization? In content? In what ways are they similar and different from the proposals in this section?

FIGURE 37-1   **STUDENT PROPOSAL FOR FEASIBILITY STUDY**

Because the proposal is short, informal, and written on campus, it is a memo.

TO: The Campus Planning Commission

FROM: Nancy Jaspers, Technical Writing student

DATE: January 15, 1988

SUBJECT: A Proposal to Study Lowering the Noise Level in the University Academic Computing Center

The subject is the noise level and the purpose is to find methods to lower the level.

This is a proposal to study the noise-level problem in the Academic Computing Center (ACC) at Pioneer State University and to find appropriate and inexpensive methods to lower this level.

Background

The sequence of events shows how the ACC has grown over the last fifteen years.

The ACC originated in 1972 when the university purchased a computer for the new computer science major. At that time, the Center contained the computer and five terminals in a 2400-square-foot area in the basement of the Administration Building. As more and more students enrolled in computer science courses, five more terminals were added. Additional space was obtained in 1978 by moving the Student Support Services to Lockridge Hall and putting the ACC in its place on the first floor. The basement then became a storage area for the ACC. Since 1978, the offices on the south side of the first floor of the Administration Building have all been turned over to the ACC.

Need

The noise level must be lowered, a convincing reason for the study.

Currently, the north side of the Administration Building houses the offices of the registrar, admissions, and payroll. Immediately above the ACC on the second floor are the offices of the president, academic dean, dean of students, and the comptroller. (See Figure 1 for floor plan.) Noise from both the computer and the printers is often overpowering and all the offices that surround the ACC have complained about excessive noise levels. Clearly, the noise level must be lowered.

North Side of Administration Building

Offices of Registrar, Admissions, and Payroll
Academic Computing Center

The First Floor of the Administration Building

North Side of Administration Building

Student Services, Counseling, and Placement Office
President, Academic Dean, Dean of Students, and Comptroller

The Second Floor of the Administration Building

FIGURE 1. LAYOUT OF THE ADMINISTRATION BUILDING

The scope specifies precisely what will be studied.	## Scope
	This proposal is to study the ways the noise level can be reduced in the ACC. The methods, if utilized, should reduce the interference heard in other offices in the Administration Building.

This section shows the methods the student will study to determine how the noise level can be lowered.	## Methods
	Four possibilities exist to reduce the noise level:
	1. install heavier ceiling tiles,
	2. rearrange the equipment,
	3. establish timetables for the use of the small printers, and/or
	4. soundproof the east-west wall on the first floor.
	Each of these methods will be studied to determine which ones are the most practical, economical, and useful in reducing the noise factor.

The methods to be explored are not costly or complicated, suggesting they could be feasible.	## Feasibility
	None of the solutions listed is extremely costly; none involves expert knowledge. Any or all of the methods may be found to be feasible. The chief factors in determining which solutions to use are listed in *Methods*.

If the ACC is quieter, everyone is happier.	## Expected Results
	If this study is accepted, the result should be a quieter ACC and a happier administration.

The student concludes by emphasizing the severity of the noise problem and the current stability in the ACC's size.	The time to act is now! The size of the ACC is fairly well stabilized, causing less concern that new facilities will be needed in the next few years. If this proposal is accepted, work could begin as soon as the final study is completed. The noise problem could be alleviated, making life more pleasant for the other occupants of the Administration Building.

Note: This is a formal proposal that needs a title page. The format for the title page was dictated by the funding institution in their Request for Proposals (RFP). Note that the amount of funding requested had to be included on the title page.

FIGURE 37-2   FACULTY PROPOSAL FOR PROJECT FUNDING

UNDERGRADUATE TEACHING IMPROVEMENT—CATEGORY A

A PROPOSAL TO PRODUCE A GROUNDWATER VIDEOTAPE
TO IMPROVE STUDENT UNDERSTANDING
OF GROUNDWATER MOVEMENT

Roger A. Nelson
Associate Professor of Civil Engineering
College of Engineering
University of Wisconsin—Platteville

TOTAL AMOUNT REQUESTED: $2296.00

Note: The abstract is at the beginning of the formal proposal, immediately after the title page. It is an informative one that summarizes the entire proposal.

A PROPOSAL TO PRODUCE A GROUNDWATER VIDEOTAPE
TO IMPROVE STUDENT UNDERSTANDING
OF GROUNDWATER MOVEMENT
ROGER A. NELSON
UW—PLATTEVILLE
DEPARTMENT OF CIVIL ENGINEERING

ABSTRACT

Groundwater is a very important resource that is becoming more polluted by man's activities on the ground surface. It is important that students in a number of disciplines have an understanding of where groundwater comes from, how groundwater moves, and how it becomes contaminated. The visualization of the movement of water under the ground's surface is difficult for many students. The underlying theoretical principles can be described mathematically, but the transition from theory to understanding in the real world is often not complete. One way of improving the understanding of the theory of groundwater movement and contamination is through the use of physical models. Groundwater movement in many cases is not rapid enough that movement in a physical model could be demonstrated in a typical lecture or laboratory period. The development of a time-lapse videotape of the model would provide a solution to both problems. The time-lapse videotape would allow for closeup viewing of the flow of groundwater in the model and a condensing in time of concepts demonstrated by the model.

The model would be built and the videotape would be produced and edited during the summer of 1986. It would be evaluated in the fall of 1986 during the normal course-evaluation procedure. Improvements to the videotape would be made after the evaluation, and the videotape would be made available to all other UW-System campuses.

Note: The RFP requested that the budget be included on a separate page at the beginning of the proposal.

BUDGET SUMMARY

NAME	TITLE	SALARY
Nelson, Roger	Associate Professor Civil Engineering	$1828.00
	Student	268.00
Supplies (1/4-in plexiglass, glass beads, aquarium sand, plastic tubing and fittings, videotapes)		200.00
	Total Requested	$2296.00

Informative title

A PROPOSAL TO PRODUCE A GROUNDWATER VIDEOTAPE TO IMPROVE STUDENT UNDERSTANDING OF GROUNDWATER MOVEMENT

Brief
introduction

Groundwater is commonly understood to refer to the water occupying the void space in some geologic formations. Utilization of groundwater dates from ancient times, although an understanding of the occurrence and movement of groundwater has come only relatively recently.

Statement of the Problem

Definition and
description of
the problem

The visualization of the movement of water down through the surface soil and into the groundwater is difficult for many students. The principles affecting the process can be described mathematically, but the transition from the theoretical to the real world is often incomplete for many students. They, therefore, have an incomplete understanding of the movement of groundwater and of contaminant movement into the groundwater. The physical factors affecting the process are complex and difficult to measure precisely. This makes it difficult to verify the mathematical equations that attempt to model the real world. Students are not easily able to assess the practical implications from the mathematics or to obtain a good grasp of the physical processes and factors at work.

Background

Discussion of a
previous attempt
to solve the
problem

One way of solving this conceptual problem would be to miniaturize the students and allow them to enter the groundwater and experience the forces at work. A second and more practical way of experiencing the forces causing the contamination and movement of groundwater is to build a physical model of the system and use this model to demonstrate the principles. A physical model of this type was presented by Margaret Blanchard of the Wisconsin State Geological and Natural History Survey at the last American Water Resources Association, Wisconsin Section, annual meeting in La Crosse, February 1985. Most of the groundwater movement shown with this model was not rapid enough to see appreciable changes in a typical class period. The effects of variations in recharge rate, dispersion of pollutants, effects of soil permeability, variations in hydraulic head, landfill leachate rate, and well pumping rate cannot all be shown at the same time, either. Each of the processes or concepts would require a three- or four-hour time period to demonstrate.

Methods

Description of a model that could be built—drawing not included in this text

The model that would provide a visualization of the principles of groundwater movement would be built of 1/4-inch plexiglass, as shown in the attached drawing. The model would be approximately twelve inches tall, twenty-four inches long, and one and one-half inches in depth. The model would be filled with sand, gravel, and bentonite clay, as shown in the attached figure.

The wells and piezometers that show the water level in each layer are made of small-diameter plastic tubing. (A piezometer is a tube that shows the water level in a given sand layer of the model.) They would be filled with red food coloring. The tubing is open only at the bottom, where it is in contact with a sand or gravel horizon, and at the top. The water in the source bottle supplies a head that forces water up into the piezometers and/or out of the flowing spring in the bottom of the lake. The landfill is made of plexiglass with perforations in the bottom. The lake is developed by gluing a plexiglass boundary, as shown, with the only inlet being the tube in the bottom that acts like a spring. Water can enter the permeable soil layers from either end by means of the inverted jar, which can be attached to either end. Water can leave through the drain in the lake or the outlet tube in the end of the model.

Objective

Subject and purpose

The objective of the project proposed here is to produce a videotape of a physical model for the situations illustrating the major principles influencing groundwater flow and groundwater movement, which are to be condensed in time and expanded in space so that the students can visualize the process. The videotape can show close up the flow pattern around soil particles and dispersion of contaminants in the groundwater.

User Group

Required section dictated by the RFP, similar to an "Expected Results" section

The initial group to benefit from this development will be students in Civil Engineering, Geosciences, Agriculture, and Reclamation at UW—Platteville. In total, the potential is for 125 students per year using the videotape at UW—Platteville. The videotape will contain an oral narrative that describes the processes occurring and the theory behind them.

The videotape will be self-contained and will not require any textbook or prior knowledge of groundwater hydrology to understand the concepts presented. The subject matter on the videotape will fit into any course that contains subject matter on groundwater quality and/or movement. All University of Wisconsin System four-year campuses offer one or more courses that could use this videotape. A professor who teaches a course that could use the videotape has been contacted at each of the campuses. All have indicated an interest in using the videotape. The number of students who would benefit from this project is conservatively estimated to be in excess of 2,000 per year.

Past Studies

Another requirement in the RFP— verification that the study is original

A review of proposals previously funded by the Undergraduate Teaching Improvement Grant program, published in *Undergraduate Teaching Improvement Grant Abstracts,* shows no previously funded proposals similar to this one.

Procedures

An explanation of what the model will demonstrate— the methods are expanded in detail

Six major concepts could be easily demonstrated with the model. These are the

1. development of a cone of depression during pumping of a well,
2. effects on water movement by impermeable layers,
3. cause of flowing wells,
4. dispersion of pollutants in groundwater,
5. landfill leachate contamination of wells, and
6. effects of soil permeability on groundwater flow rates.

Each of these six concepts would be demonstrated by adding food coloring to the part of the groundwater that is moving the most rapidly or to the piezometers that show changes of water level in the soil. For the first demonstration, food coloring is added to the piezometers, and the wells shown in the attached figure are pumped one at a time. The water level is drawn down in the piezometers in the same soil layer as the well and near the well.

The bentonite clay layers act as boundaries that keep water from moving vertically in the model because they are much less permeable than the overlying and underlying material.

A flowing well is demonstrated by lowering the water level in the lake from what is shown in the figure and maintaining the water in the inlet zone as high as is shown. The tube that has a removable connection near the bottom of the lake is disconnected and water can flow into the bottom of the lake.

The dispersion of pollutants is demonstrated by pumping food coloring down one of the wells and observing the spread of the color as it moved into the sand. The landfill leachate contamination of wells is demonstrated by adding food coloring to the landfill and pumping the nearest well.

The effects of soil permeability is demonstrated by replacing the sand with gravel and repeating any one of the demonstrations listed. A comparative value for permeability could be obtained by placing fine sand in the upper layer, coarse sand in the middle layer, and gravel in the lower layer, and introducing food coloring with the water in the inlet. The food coloring travels more rapidly in the coarser material.

After the filming of each of the demonstrations outlined here and the videotape has been edited/ condensed in time, a narrative description of the process would be added. The narrative would describe the theory involved in each demonstration.

Project Detail

Another required section—note change to imperative mood (implied "you")

1. Build the model.
2. Write the narrative for each demonstration.
3. Perform each of the demonstrations and film them.
4. Edit the videotape.
5. Add the oral narrative to the edited videotape.

Personnel

An additional two-page résumé is not included in this text

Dr. Nelson teaches courses that include a discussion of groundwater movement and contamination. He has also obtained detailed information from Margaret Blanchard of the Wisconsin Geological Survey in Madison on the construction of the physical model described and shown in the attached figure. Dr. Nelson would build the model and write the narrative for the demonstrations described above. A student from the UW—Platteville Department of Communications with a knowledge of filming and editing a videotape would be employed to complete those parts of the project. The Director of TV Services at UW—Platteville has been contacted to assist in the filming of the experiments. He has indicated a willingness to assist on the project and will make available the video camera and editing facilities at UW—Platteville.

Evaluation

A final requirement in the RFP

Class evaluations at the end of the semester will be used to evaluate the success of the inclusion of the videotape. Specific questions will supplement the usual course evaluation form. These questions will address the enhancement of understanding of basic principles as a result of the videotape presentation. Copies of the videotape will be made available to faculty at other universities for their evaluation and review. Their comments will be incorporated in the improvements made to the model and videotape. Upon completion of the project, a paper describing the use of the videotape of the groundwater model will be submitted to an appropriate refereed journal for publication.

●38

ORAL REPORTS

KEY POINTS

- Prepare thoroughly for an oral report.
- Practice giving an oral report beforehand.
- Prepare readable visual aids.
- Speak clearly and distinctly.
- Finish within the allotted time.

An oral report is similar to a written report with a few notable exceptions. It has an introduction, body, and a terminal or closing section just like a written report. But an oral presentation has some advantages over a written report. It is more personal and provides immediate feedback from the listener. If presented well, an oral report is more memorable and less time consuming for the audience compared to reading a report.

Oral reports also have the following disadvantages when compared to written reports:

1. They are not as easy for the preparer to organize and polish.
2. The presentation of the technical information must be less complex.
3. The graphics used to present data often must be simplified from those found in a written source.
4. The listener has no opportunity to go back and look at the material again.

38.1

Oral presentations made before small informal groups, co-workers, or superiors are very similar to large formal presentations. The choice of visual aids might be different for one type of presentation than for the other because of the makeup of the group. The preparation steps for both groups are the same, however.

This section contains information on how to prepare for giving oral reports, how to present an effective oral report, and how to follow up at the end of an oral report with a question-and-answer period. It is not intended to substitute for a formal course in public speaking, but it does provide some guidelines for making an effective presentation of technical information.

38.1 PLAN THOROUGHLY FOR AN ORAL PRESENTATION.

In preparing for an oral presentation keep two points in mind. First, always analyze your audience. The purpose of your presentation is to communicate information effectively. In order to do this you must know the background of your audience and coordinate the level of your presentation with the knowledge your audience has of your topic. Do not insult your audience's intelligence by providing background information and definitions of terms that are common knowledge to them. For example, if you are speaking on new developments in electronics before a group of electricians, do not attempt to define electrical terms that the technicians would not only be familiar with but would use every day. Likewise, do not present information that is above the level of understanding of your audience so that it is incomprehensible or lost after a very short time. For example, the same talk on new developments in electronics would need to be simplified and terms would need to be defined if you were presenting it to your local Kiwanis Club where, quite possibly, no one has much knowledge of electricity.

If your topic is a controversial one, you need to know the prejudices of the audience. It is important not to offend anyone, particularly if you are seeking their support on a position.

The second concern is the facility where your oral report will be given. You need to know how large the room is, what visual-aid equipment can be provided, and whether a loud-speaker system is available to use during the presentation. Plan your presentation to prevent surprises about the facility that would be distracting and cause a less than effective oral report.

An effective oral presentation also requires adequate preparation and knowledge of the material. Your preparation for an oral report involves several steps. First obtain and assemble the information you are going to present.

38.2   PREPARE AN OUTLINE.

Once you have assembled the information, your next step is to prepare an outline—the framework around which the presentation is organized. This helps ensure that you consider all aspects of your presentation. The outline for an oral presentation is similar to that for any speech and contains an introduction, a body, and a closing.

a. Introduction

Your introduction should have a strong main-subject statement that gives your audience an idea of what your talk is about. It tells them the focus for the remainder of your presentation. If your presentation is to effectively inform the audience, in your first few sentences you must capture their attention and arouse curiosity about your topic.

b. Body

After the introduction, the next section is the body of the report. This is equivalent to the body of a written report. For example, if you are reporting on a study, begin with the approach you took in your investigation, proceed with the subject of your report, and finish with your objectives. Think about what your audience is to remember from your report and plan how best to achieve that. The organization of the presentation is much more important in an oral report than in a written report because the audience will only see and hear it once. Select visual aids that help the listener to understand and remember each point you make.

Provide smooth transitions between the introduction and the body of the report and between the specific points made in the results sections. In a written report subject headings set off each section and provide organization to the report. These are not present in an oral report, but you can substitute them with clear transitional sentences. For example, you can use several sentences to introduce each point presented in a sequence. Discuss each point completely before going on to the next point.

c. Conclusions

The third section of an oral report is the terminal section. This section is also similar to the conclusions in a written report. It contains a summation of the results' discussion, recommendations based on the conclusions, and a closing statement.

Your summary reinforces the main points you made in the discussion of the results section. At the beginning you tell the audience what the report is about; then you tell them your information. Finally, you tell them what they have been told. This repetition of information is not as necessary in a written report because the reader can always go back and look at the report again. Once an oral report is over, the listener has only a memory of what was said. To help the audience remember the important points, repeat them in the summary.

You may find it appropriate to recommend that some action be taken as a result of your investigation. Present this after you state your conclusions. Your recommendations may be to take certain steps or to follow up on the study with a more in-depth investigation of some particular aspect you have reported. Then provide a closing statement to signal you are finished.

38.3 PREPARE EFFECTIVE VISUAL AIDS.

After you have prepared the outline of the report and have identified the main points to be made, you are ready to select the appropriate visual aids to accompany your presentation. Either check out beforehand the facilities where your presentation is to be made or, if you cannot visit them yourself, contact someone who can to ensure that the visual aids will be readable to everyone in the audience. Select the appropriate visual aids based on how large the audience is, on what aids would help clarify the presentation, on the shape and lighting of the room, and on the projection equipment available. Slides and transparencies will work for just about any size room and audience. Chalkboard drawings, blueprints, engineering drawings, and posters are only effective for small groups.

The visual aids should support the points you make in the presentation of and discussion of the results. Slides and transparencies are the most common visual aids for oral presentations. Slides are used to best advantage where pictures of particular objects and the use of color enhance a presentation. Clear plastic transparencies have advantages over slides in that the speaker can face the audience, point out the main features directly on the transparency, and talk to the audience all at the same time.

a. Slides

Arrange your slides in a specific order before your presentation, for it is difficult to go back and forth between several slides without showing those in-between and distracting the audience. If it is necessary to show the information in one slide more than once, make multiple copies of it and place them in the sequence where they are needed.

b. Transparencies

Use transparencies for graphs, charts, and diagrams. You can write on clear plastic sheets during the talk but it is much better if you prepare transparencies well ahead of time. You can place transparencies on top of each other during the presentation to show how one point or feature is related to another. You also can draw lines or other symbols on the transparency in color while the transparency is being displayed. Transparencies can be shown in any sequence. Be sure to arrange your transparencies in the sequence you want to present them to prevent the distraction of having to shuffle through the sheets to find the appropriate one.

Several points about transparencies should be remembered. First, static electricity tends to make them stick together unless you place a cardboard border on each one; label the border to help you find particular transparencies. A thin paper spacer also can help you overcome static electricity and separate the transparencies from each other. The spacers make it difficult to find a particular transparency, though, if the transparencies get out of sequence.

Simple visual aids are best. Place only one point or a few points on each graphic. Graphics developed for written reports usually con-

38.4

tain too much information and need to be modified for oral presentations. The reader of a written report has time to study a table or figure but a viewer of an oral presentation only has the brief time that a visual aid is on the screen. Include only enough information on each visual aid to make the point.

Know which type of graphic to use. For example, line or bar graphs show relationships and trends much better than do tables, making comparisons easier to visualize. For further discussion of Graphics, see Section 39.

38.4 MAKE AN EFFECTIVE PRESENTATION OF AN ORAL REPORT.

a. Practice

The best way to make an effective oral report is to be well prepared by practicing the presentation many times. This allows you to become thoroughly familiar with the material and to develop confidence in its delivery. The extemporaneous speaking technique is best for an oral report. Follow the outline to prepare your notes. Write only key points or phrases in your notes, not entire sentences. You can place your notes, typed or handwritten, on note cards; typed on typing paper; or handwritten in a spiral notebook. A notebook would keep your notes from getting mixed up. If your notes are on cards and the cards are dropped during your presentation, you may never find your place again. To avoid the latter, you can number the cards so that you can quickly put them back in order, but the interruption may ruin your presentation.

It is helpful to practice the presentation with the notes in front of an audience of coworkers or friends. If you stumble over or forget some points, you can revise your notes and repeat the process until your delivery is smooth and within the allotted time. After you have practiced your presentation a number of times, you may find that your notes are no longer necessary and the visual aids themselves are a reminder of what to say next. If no practice audience is available or even if one is available, videotaping the presentation and viewing the tape allow you to see and hear where improvements can be made.

b. Enthusiasm

Several subtle features of the delivery affect the way a presentation is received by your audience. Show enthusiasm for your subject. You do this more by the way in which you speak your words than by the actual words. You can make your talk more interesting by mentioning personal anecdotes about your involvement in collecting the data or samples. Select with care stories that show your personal interest in the subject. Your audience will be convinced of your conclusions by the strength of the evidence you present, not by the humorous stories that accompany the talk.

In addition, the speaker can control a number of other characteristics of a good oral presentation. Keep these points in mind:

- speak with confidence and authority, but don't overstep the bounds of common courtesy
- appear to consider each member of the audience an equal
- do not speak down to the audience
- do not apologize for the talk or visual aids
- establish eye contact with as many members of the audience as possible
- do not use multisyllabic words for effect

The audience will quickly lose interest if you use too complex a vocabulary and unknown technical terms. One noted university professor had the habit of frequently using multisyllabic words. His students quickly lost interest in the class subject and resorted to playing "buzzword bingo." They kept track of which big words he used and how often he used them on each day of the semester.

Your appearance and personal mannerisms also affect how your presentation is accepted by the audience. Avoid distracting habits like playing with keys or coins, pulling on your hair, or tugging at your clothes. Be sure to wear attire appropriate for an oral report and the audience you are addressing.

Nervousness at the beginning of an oral report often causes a person's speech to speed up. The pace of your presentation must be slow enough so that all words are pronounced clearly. Speaking at a rate that seems too slow to you is probably about right for the audience.

38.5 BE PREPARED TO ANSWER QUESTIONS.

At the conclusion of the oral report a compliment is often made to the audience thanking them for their attention. It is also appropriate at this point to ask if anyone has questions. If the report is at a formal conference, the moderator usually performs this function.

In the beginning stages of assembling the information for your report, you prepared more than you had time to present. Such a thorough study of the subject can help you to anticipate what questions you might be asked and to prepare answers for them. Adequate preparation makes you more knowledgeable about the subject than anyone who is expected in the audience.

Despite a thorough preparation, you still may not be able to answer some questions. Do not try to change a question into something that can be answered. Simply state that you do not know the answer to the question, which is better than trying to bluff the questioner.

If you present your oral report at a large conference with a session moderator, the moderator may restate the question to ensure that everyone heard it. If no moderator is present, you should restate the question before answering. This ensures that everyone heard it and allows you time to phrase an answer. Be careful with your tone of voice and choice of words in answering questions to avoid any confrontations with those who pose questions and to treat the audience with respect.

Most conferences schedule presentations for twenty to thirty minutes. In consideration of the other speakers, each speaker should finish within the allotted time. To keep the program on schedule, the moderator or speaker may need to cut off questions. One way of limiting questions is by stating at the beginning of the question-and-answer period that there is time for only a specified number of questions. Conferences often have multiple sessions occurring simultaneously. Speakers who stay on schedule allow the audience to switch from one session to another between speakers. This also ensures that all speakers have enough time to finish their presentations. If the moderator signals that time is up, do not say, "I'll just borrow a little time from the next speaker." Instead, conclude with a comment that if anyone is interested in pursuing the topic further, they can see you after the session.

Time the presentation during the practice sessions and adjust it to fit within the allotted time. By practicing and being prepared for the oral report, you should have no problem with running over the allotted time.

38.6   PREPARE A WRITTEN REPORT TO DISTRIBUTE.

Write a formal report to accompany oral presentations at regional or national meetings. The written report can include details that are too lengthy or complex to discuss orally. When you discuss the main points of a study in the oral presentation, make reference to the more detailed discussion in the written report. Do not hand out the written report until after you have finished your oral presentation; announce at the beginning that copies of the report will be available afterward.

●39

GRAPHICS

KEY POINTS

- Keep each graphic as simple as possible.
- Give each graphic a clear, concise title.
- Mention each graphic at least once in the text.
- Place each graphic immediately after it is first mentioned in the text.
- Explain all symbols and abbreviations used in graphics.
- Orient all graphics the same direction.
- Use the appropriate graphic for the information in the report.

Graphics include tables, charts, graphs, or other illustrations used in technical writing to supplement quantitative or pictorial written information. The use of appropriate graphics makes mathematical or conceptual relationships easier to understand. Combined with explanatory text, they can clarify a point or concept better than hundreds of words of text.

Since the purpose of technical writing is effective communication, graphics should be used only when necessary to help readers understand the information. They are not intended to just make a communication look more attractive or more complex.

No inconsistencies should exist between the graphics and the text because they make the report's conclusions suspect. For example, a table may indicate that pipe costs $25 per foot, but if the text that accompanies the table states that pipe costs $30 per foot, the reader does not know what to believe.

Graphics in this section are divided into the following categories: tables, charts, graphs, illustrations, and computer graphics.

Some general rules apply to the effective use of all types of graphics:

1. Keep each graphic as simple, clear, and logical as possible.
2. Number each graphic by order of appearance and give it the appropriate heading: label as a figure everything that is not a table.
3. Give each graphic a clear and concise title.
4. Mention each graphic at least once in the text.
5. Place the graphic close to and following the first mention of it in the text.
6. If necessary, use footnotes to explain or clarify entries in graphics.
7. Surround each graphic with enough white space to set it off from the text.
8. Be sure all graphics are self-sufficient for reader comprehension.
9. Acknowledge the sources of all information at the bottom of the graphic.
10. Clearly explain all symbols and abbreviations used in graphics.
11. Keep the margins of the graphics within the margin of the text.
12. Orient all graphics the same direction as the text or with the bottom parallel with the right margin of the page.

39.1   CONSTRUCT TABLES CLEARLY AND ACCURATELY.

Use tables to display related items in rows and columns. The row items may be either numbers or words. Because some related items are repetitive, a table reduces the amount of text that would be required to present the same information and make comparison of the items easier.

Tables may be either formal, informal, or a matrix. See *Section 30* for details about informal tables.

a. Formal tables

Formal tables are set off from the rest of the text. They require table numbers and titles and have row and column headings.

The following rules apply to all formal tables:

1. Express fractions as decimals unless this is not customary in a particular technical field or company or gives a false sense of precision.
2. Align all digits so that the decimal points are in a vertical line (see Section 12.9, p. 83, for a more complete discussion).
3. Use horizontal lines to separate the column headings from the numbers or words in the table and after the last row.

Note: Solid lines should not be used to form boxes around each item, except in a matrix (see Figure 39-2).

4. Place like items in a table in columns rather than in rows so that the table takes up less space and, more importantly, so that a comparison between items is easier for the reader.
5. Place units of measure, amounts, or any other symbols in the column and row headings, not in the body of the table. The only exception to this occurs when the units vary within a column or row.
6. Use parallel grammatical form for all headings.

The table in Figure 39-1 has specific column headings: *Wet Tons*, (*Wet Metric Tons*), and so on. The row headings are listed under the column heading *Component*. Parallel grammatical form has been used for all headings. Units of measure are specified in the column headings and do not appear in the individual items in the body of the table.

b. Matrices

A matrix is a type of table that contains headings for the rows and columns, and lines between each row and column. Words, numbers, or symbols are placed in the boxes created by the lines. The symbols are used to indicate whether an item is or is not present at the intersection of a row and column. The campground list in Figure 39-2 is an example of a matrix.

FIGURE 39-1   EXAMPLE OF A TABLE

MATERIALS BALANCE FOR COMPOSTING OPERATION

Component	WEIGHT			DENSITY		VOLUME		Percent Volatile Solids
	Wet Tons	(Wet Metric Tons)	Percent Solids	lb/yd³	(kg/m³)	yd³	(m³)	
Undigested Dewatered Sludge	70	(64)	20	1,600	(950)	88	(67)	75
New Wood Chips	13	(12)	70	500	(295)	52	(40)	90
Recycled Wood Chips	39	(35)	70	600	(355)	131	(100)	80
Mix	123	(112)	41	900	(535)	271[a]	(207)	80
Wood Chip Pad	9	(8)	70	500	(295)	34	(26)	90
Unscreened Compost Cover	19	(17)	65	725	(430)	51	(39)	65
21-Day Compost Pile	90	(82)	65	725	(430)	248	(190)	65
Screened Compost	32	(29)	60	975	(560)	66	(51)	45

[a] Although this hypothetical mix volume is presented here as the sum of the volumes of the component parts (dewatered sludge, new wood chips, and recycled chips), in reality the mix volume will be significantly less than the sum of the parts.

Source: Singley, M.E., A.J. Higgins, and M. Frankin-Rosengaus, *Sludge Composting and Utilization: A Design and Operating Manual*, 1982, New Jersey Agricultural Experiment Station, Rutgers University, New Brunswick, N.J.

FIGURE 39-2 EXAMPLE OF A MATRIX

WISCONSIN STATE PARK INFORMATION

STATE PARKS	NEAREST COMMUNITY	INDEX	Number of Camping Units	Electric Outlets	Flush Toilets	Showers	Fishing	Nature Trails	Hiking	Swimming	Picnicing	Cross-Country Skiing	Canoeing	Boating	Concessions	Nature Center
Amnicon Falls	Popular	2C	40	X	X			X	X		X					
Belmont Mound	Belmont	10E						X	X			X				
Brunet Island	Cornell	5D	69		X	X	X	X	X				X	X	X	X
Devils Lake	Baraboo	9F	459	X	X	X	X	X	X	X	X	X	X	X	X	X
High Cliff	Sherwood	7H	54					X	X		X	X				
Mirror Lake	Delton	9F	144	X	X		X			X	X		X	X		X
Peninsula	Fish Creek	5J	13		X	X	X	X	X			X				
Potawatomi	Sturgeon Bay	6J	125	X	X		X	X	X	X	X			X	X	
Yellowstone	Blanchardville	10D	129	X			X	X	X	X	X	X	X	X		X

Source: Wisconsin State Highway Map

In the campground list the main column headings are the facilities available at state parks, forests, and trails. Each of the column headings—showers, concessions, and so on—intersect with the row headings—the state parks, forests, and trails.

39.2   CONSTRUCT AND USE CHARTS CORRECTLY.

Three types of charts are frequently used in technical writing: the pie chart, the organizational chart, and the flow chart.

a. Pie charts

Use pie charts to compare parts of a whole with each other and as a whole. They are applied most frequently in business and economic reports, because the reader, regardless of his or her technical background, is able to see and understand the relative size of the parts of a budget or expenditure.

The following are general rules for constructing pie charts:

1. Begin the largest wedge or percentage at the top of the pie or in the twelve o'clock position.
2. Proceed clockwise with the next largest percentage.
3. Use no fewer than four and no more than ten wedges in the pie. Too few wedges give little to compare. Too many wedges can be confusing.
4. Combine small segments into one and label it "Other." Place a note at the side or bottom of the chart to indicate what is in the "Other" category.
5. Label the wedges horizontally on the page. Labels may be either inside or outside of the wedges as long as you clearly identify your wedges and the graphic is appealing.

The pie-chart wedges often contain both a percentage and the actual quantity. In constructing a pie chart, begin by dividing it into percentages. Obtain the correct angle for the borders of each wedge by multiplying the percentage of the whole that the part comprises by 360, the number of degrees in a circle. Thus, each percentage point is worth 1 percent of the circle or 3.6 degrees. For example, 10 percent would comprise 36 degrees; 20 percent, 72 degrees; and 50 percent, 180 degrees, of the circle.

FIGURE 39-3 EXAMPLE OF A PIE CHART

Note: Pies included in the *Other* category are mincemeat, chocolate, and coconut cream.

Once you have established the size of the wedges, you can draw the borders on the pie and add the labels.

Figure 39-3 is a typical pie chart. It shows the types of pies sold at Dick's grocery store. The most popular pie is apple, 29 percent of the total. The apple pie wedge starts at the 12 o'clock position. The second most popular pie is cherry at 22 percent, which follows after apple clockwise around the pie chart. All other types of pies, after the five most popular, are combined into the "Other" category with 10 percent.

b. Organizational charts

An organizational chart is a graphic for displaying the chain of command within an organization. Figure 39-4 is an example of an organizational chart for a public works department. The titles for each of the positions are indicated in boxes that are connected to each other with lines. The highest level of authority is at the top: the Water/Sewer Commission and the Common Council. Next in the chain of command is the City Manager. Below the City Manager is the Director of Public Works. The other job classifications are shown below their respective supervisors.

FIGURE 39-4   EXAMPLE OF AN ORGANIZATIONAL CHART

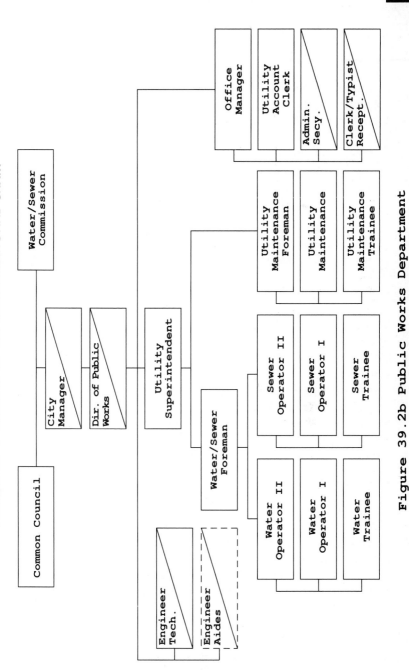

Figure 39.2b Public Works Department

c. Flow charts

Flow charts are used to illustrate functional relationships in a process or organization. Figure 39-5 is an example of a flow chart for a wastewater treatment plant. The individual processes are represented in figures similar in shape to the actual units. The processes are connected with broken and solid lines to indicate piping and water-flow. The direction of flow is indicated by arrows along the lines or inside the process units.

39.3   PLOT LINE AND BAR GRAPHS ACCURATELY.

Graphs are used to show such numerical data as trends, distributions, and cycles. A table displays information more accurately than a graph, but the relationships and trends are harder to comprehend quickly.

A graph is a plot of a set of ordered pairs or coordinates. The set of ordered pairs describes a relationship between the first element of the pair and the second. The elements of the pairs are usually numbers but often the horizontal coordinate states a category and the vertical coordinate states a unit of measure. One element of the pair is called the *dependent variable* and the other is called the *independent variable.* Examples of independent variables are time, temperature, and velocity. The independent variable is usually plotted on the horizontal or x-axis of a graph and is referred to as the *abscissa.* The dependent variable is referred to as the *ordinate* and is usually plotted on the vertical or y-axis.

A graph may be a line graph, bar graph, surface chart, histogram, or pictograph.

a. Line graphs

A line graph is a plot of coordinates of numerical data on paper that has a rectangular grid. The data points are measurements or values in the relationship between the dependent and independent variables. The relationship is a continuous one, and the points are connected with a line. Often a smooth curved line is drawn between the points to show the general trend of the relationship. The individual points are sometimes omitted and only the line is drawn on the graph to show the relationship. The choice of including or omitting

FIGURE 39-5   EXAMPLE OF A FLOW CHART

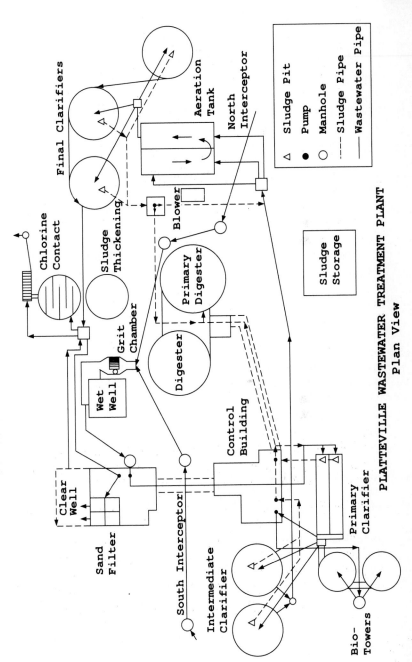

PLATTEVILLE WASTEWATER TREATMENT PLANT
Plan View

the points depends on your needs. If you need to know only the general relationship, leave the data points off.

The graph's scale must be a correct visual representation of the relationship you are presenting. The slope of the line indicates the rate of change of one value relative to the other. If the slope is steep, the reader may think that it is very significant. Therefore, the choice of scale for the x-axis and y-axis may make the slopes appear more or less significant than they are.

You can use one plot to compare several dependent variables that are a function of one independent variable. Combining solid, dashed, and dotted lines, or using different colors for each line makes it easy to distinguish them; however, if the graphs are to be photocopied, do not use colored lines. Not only is the photocopy black and white, but some colors do not photocopy well.

Enumerate the units of the ordinate and abscissa only for the range of data plotted. For example, the axis should not continue to 100 if the largest data point is only 72. For readability, use axis lengths that are easily divisible by two, five, or ten units. If the axis were divided up into multiples of twenty-three, for example, the reader would have difficulty finding a value for nine. Choose subdivisions along the axis that separate the numbers from each other but that are not so small as to be unreadable. The axes usually start at the coordinates (0,0), but you determine the starting values by the data to be displayed on the graph.

If your graph is meant to show only general relationships, omit the grid lines. Grid lines make a graph more cluttered and the relationships harder to visualize. When the reader needs to estimate a value between plotted points or to read precise values from the graph, include the grid lines.

The following are some general rules for constructing line graphs:

1. Include the data points on the graph only if you need exact values.
2. Use the largest grid size that is compatible with the data.
3. Use different symbols for the points or different types of lines for each set of data when more than one dependent variable is included.
4. Label each line or include a key or legend for each line of the graph.
5. Label the axes with the appropriate numbers.

FIGURE 39-6   EXAMPLE OF A LINE GRAPH

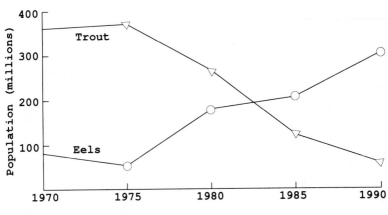

Figure 39.3a   Effect of eels on trout

Figure 39-6 is an example of a line graph showing the population of trout and eels from 1970 to 1990. As the eel population increased the trout population decreased. Both populations are indicated with solid lines but a different point symbol is used for each. The lines are also labelled for ease of understanding.

b. Bar graphs

A bar graph presents information that is to be compared to only one variable or axis. Draw bars to indicate the size of the variable for each category displayed. The longer the bar, the greater the quantity being indicated. You can easily compare the magnitude of the categories.

Place bars horizontally or vertically, depending on the desired effect. Vertical bars are used for variables like height or dollars. Horizontal bars are used for variables like distance and time.

The following are some general rules for constructing bar graphs:

1. Draw all bars with the same width, both for uniformity and to present a balanced visual for the reader.
2. Place the same amount of space between the bars for balance and uniformity.
3. Limit the amount of space between the bars to no more than the width of the bars themselves.

319

4. Use sufficient space between the bars to accent the differences between the categories.
5. To show divisions within each category, divide each bar into no more than three subdivisions. Either label these subdivisions on the bars or use a key in the graph or a legend at the bottom of the graph to explain each bar.
6. Draw the bar graph to an appropriate scale. The visual comparison of the bars should be the same as the values in a numerical comparison; the length of the bars on paper should represent their true numerical value.
7. If the bar is too long for your graph, clearly indicate a break in the scale, but don't allow the scale break to distort the relative magnitude of the bars.
8. Do not distort a numerical comparison in which one value is of greater value than another. For example, if one value is three times the other, do not choose a scale that makes the respective bar appear on the graph only two times greater than the other.

Figure 39-7 is an example of a bar graph. It displays the mineral element content of forest floor litter in kg/ha. Each element has its own bar and the relative magnitude of the elements can be seen.

FIGURE 39-7   EXAMPLE OF A BAR GRAPH

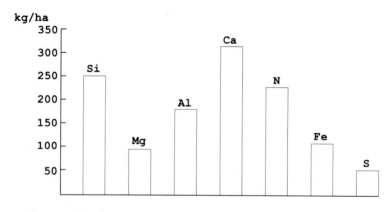

Figure 39.3b Mineral element content of forest floor litter

c. Surface charts

A surface chart is a graph showing the change in several similar variables with respect to a common parameter like time. The chart is like a line graph in that changes in each variable are shown by a line on a grid. It differs from a line graph in that the quantities are additive and the top line shows the total. In Figure 39-8, the energy consumption by various sources from 1971 to the year 2000 is shown. Energy consumption is broken down into five categories: hydropower and geothermal, natural gas, petroleum, nuclear power, and coal. The amount of natural gas is added to the amount of hydropower and geothermal. The line at the top of natural gas represents the sum of both, and the line above coal shows the sum of all five.

Use this type of graph when you want to show the amounts of each part of a whole as a function of some other variable like time. A surface chart also provides a relative measure of the contribution of each to the total. It is similar to the type of bar chart in which the

FIGURE 39-8   EXAMPLE OF A SURFACE CHART

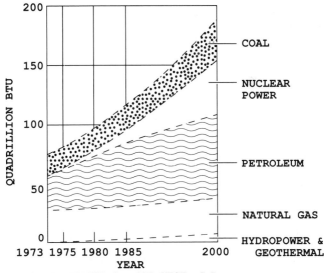

SOURCE: DUPREE, WALTER G., AND WEST, J.R.,
           UNITED STATES ENERGY THROUGH THE YEAR 2000,
           DEPARTMENT OF THE INTERIOR, DECEMBER 1972.

Figure 39.3c Energy Consumption

FIGURE 39-9 EXAMPLE OF A BAR GRAPH

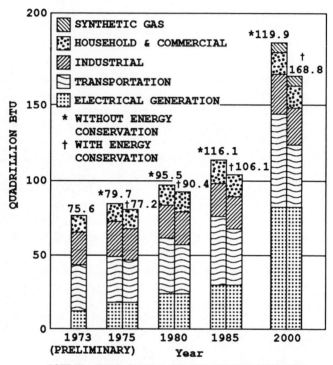

SOURCE: USDI OFFICE OF ENERGY CONSERVATION

Figure 39.3d Energy Consumption

bars are broken up into components, as in Figure 39-9. This chart shows the breakdown of energy consumption at discrete points in time instead of the continuous change that some surface charts show. Shade the components of a surface chart to set off one part from the ones plotted next to it.

d. Histograms

A histogram is a type of graph used to indicate the frequency of occurrence of a particular item. Histograms look like bar graphs without any spaces between the bars. They are used most commonly to present statistical information for business and engineering applications. Figure 39-10 is an example of a histogram. It shows the fre-

FIGURE 39-10   EXAMPLE OF A HISTOGRAM

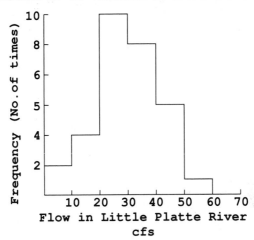

quency of flood peaks in the Little Platte River. Stream flow is a continuous variable that has been divided into class intervals of 100 cfs for display in the histogram.

The same rules for labeling line-graph axes apply to histograms, surface charts, and bar graphs.

e. Pictographs

A pictograph is a bar graph in which the bars are made of a string of items that the bar represents, such as fish, coins, people, or ice cream cones. Each item represents a specific quantity. Using pictures is eye-catching and helpful for a lay audience. The length of the bar and not the size of a pictured item represents the magnitude of the category being displayed. All items are kept the same size so that the true relationship between the categories is preserved. Figure 39-11 is an example of a pictograph.

39.4   USE ILLUSTRATIONS IN A REPORT TO CLARIFY MATERIAL.

Illustrations include diagrams, photographs, drawings, and maps. The purpose of illustrations, as with other graphics, is to clarify the material presented in the text of a report.

FIGURE 39-11 EXAMPLE OF A PICTOGRAPH

Cast Iron	$ $ $ $ $ $ $ $ $ $ $ $ $ $ $ $ $ $ $
Vitrified Clay	$ $ $ $ $ $ $ $ $ $ $
Concrete	$ $ $ $ $ $ $ $ $ $ $ $ $ $ $
Plastic	$ $

Total Sales of Pipe
Each $ = One Million Dollars

a. Diagrams

Diagrams are graphics used to depict organizational or functional relationships. You must consider the detail of the diagrams carefully, as well as what the reader needs to know, before deciding whether to include them. For example, detailed electrical-circuit schematics may be more confusing than helpful to readers. Instead, a simple, functional line drawing may be more appropriate in the body of the report. The detailed schematic can be included in the appendix, if necessary.

In Figure 39-12, a functional diagram illustrates the relationship between the water zones below ground. The components of the saturated and unsaturated zones are clearly defined and labelled.

Figure 39-13, p. 326, is an organizational diagram of the computer facilities at Pioneer University. As well as the organization of the facilities, it illustrates the functional relationship among the individual components of the system. The main computers are identified and the ethernet (network lines) that links them is shown. The terminals for access to the computers are labelled, and the line printers used to obtain output from the computers are identified.

b. Photographs

Photographs are used for clarification of objects or places. They provide a clear definition of minute details to separate or identify

FIGURE 39-12   EXAMPLE OF A FUNCTIONAL DIAGRAM

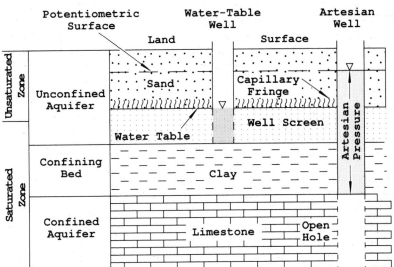

Aquifiers and Confining Bed
Source: L. Lawrence Graham et al., *Protection of Public
        Water Supplies from Ground-Water Contamination*,
        Seminar Publication, EPA Technology Transfer,
        EPA/625/4-85/016, Washington, D.C.

similar items. Photographs furnish much better proof of a point than any amount of text. However, be careful when selecting them for reports. They may contain much extraneous information that might confuse more than clarify the point being made in the text.

One way to indicate important points is to add labels or arrows to photographs. In photographs of objects, a sense of scale must be present. You can provide this by placing a ruler or a familiar object, such as a coin, in the photograph before you take it. Another method is to draw a scale on the bottom of the photograph itself.

While photographs may provide an excellent graphic display to aid the reader, they are expensive to reproduce for multiple copies of a report. They also are difficult to duplicate on photocopiers with the same level of detail as the original.

c. Drawings

Drawings often illustrate the details of an object, mechanical device, or place better than photographs because they can reveal many

39.4

FIGURE 39-13 EXAMPLE OF AN ORGANIZATIONAL DIAGRAM

Computer Facilities at Pioneer State University

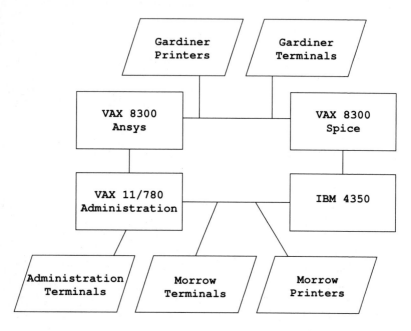

angles, close-ups, cutaways, and the like. For example, drawings from different perspectives can show three-dimensional characteristics and details. Portions of an object can be cut away to indicate more detail or its internal structure and workings. Exploded views also can show the inside pieces of an object. Portions of a drawing that contain many small parts can be magnified to better illustrate the relationships among them.

Figure 39-14 is a drawing of a sediment incubation reactor. All parts of the reactor are labelled and the interior mixing and aeration system is illustrated to aid the reader in understanding how the system operates.

d. Maps

A map is a graphic representation of features on the earth's surface plotted to a definite scale. The features may be manmade, natural, or both. Use maps to show locations where samples were collected, the geographic distribution of something, or geologic features.

FIGURE 39-14   EXAMPLE OF A DRAWING

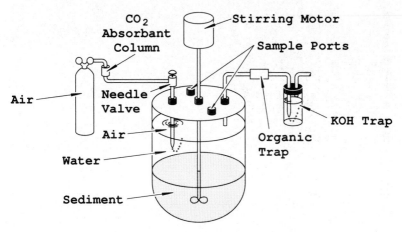

Sediment Incubation Reactor

Be sure your map contains a scale in appropriate conventional units, a "north" arrow, and a key or legend that identifies the symbols used. The information you show on a map should be easy to comprehend and crucial to the reader's understanding. Simple maps that demonstrate only the points being made in the report are the most easily understood. If you are indicating the geographic distribution of several items, use a different type of shading for each item to make it easily distinguishable from the others.

Figure 39-15 is a map of the sampling locations in a study of the atmospheric fallout of PCBs into Saginaw Bay. The location of Saginaw Bay in Lake Huron, the larger body of water of which it is a part, is shown in an inset map.

39.5   USE COMPUTER GRAPHICS.

Computer graphics are tables, graphs, charts, or other illustrations that are created with a computer. The graphics can be selected from a display on a terminal or television-like screen in as many colors as desired. The data used to produce the graphic can be altered and the figure redrawn in a matter of seconds or minutes. The size of the graphic can be altered easily by changing a scale factor. Some programs, such as Lotus 1-2-3, can display the information as

FIGURE 39-15 EXAMPLE OF A MAP

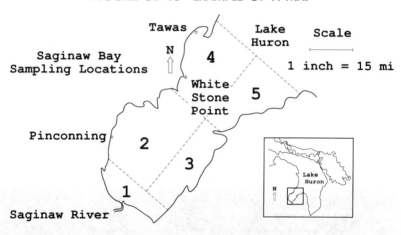

either a pie chart, a bar graph, or a line graph. The graphic that best displays the information and supports the point being made then can be drawn by computer plotter or graphics printer.

Figure 39-16 shows a grain size analysis of a soil sample displayed in all three Lotus 1-2-3 graphic forms. The most appropriate one for the analysis would be selected for the report.

It may take just as long to generate the first version of a graphic with a computer as to draw it by hand. However, the advantage of computer-generated graphics over hand-drawn graphics is their appearance and the ease and speed of changing a drawing once the data have been entered into the computer. The scale can be changed easily and the graphic redrawn until the desired effect has been achieved.

Computer-generated engineering drawings are also becoming more and more common. In addition to producing the final engineering drawing, computers are used to design and analyze a product or structure before it is built. Computer graphics and analysis systems allow this design process to go from concept to completion more efficiently. The computer can generate engineering drawings that show a three-dimensional object. The object can be rotated, and a perspective or plan view can be made of any side.

Figure 39-17, p. 330, is a computer-drawn three-dimensional object viewed from different sides.

FIGURE 39-16   EXAMPLE OF COMPUTER GRAPHIC OPTIONS OF LOTUS 1-2-3

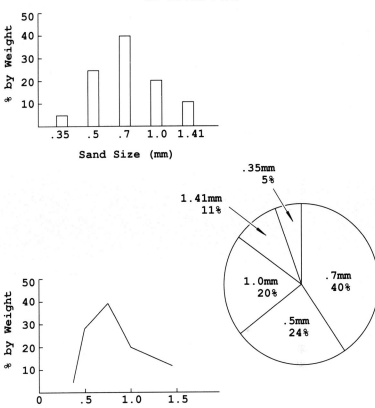

● EXERCISES

Be sure to provide the appropriate title, label, legend, key, figure number, or table number for each exercise.

1. Visit the local water or wastewater treatment facilities and devise a flow chart for the process used. Provide a brief discussion of the process.
2. Obtain records of the daily temperature and precipitation for the last month. Display this information in the appropriate graphic and discuss the data with respect to the average daily values for the time period.

39.5

FIGURE 39-17   EXAMPLE OF A COMPUTER-DRAWN THREE-DIMENSIONAL OBJECT

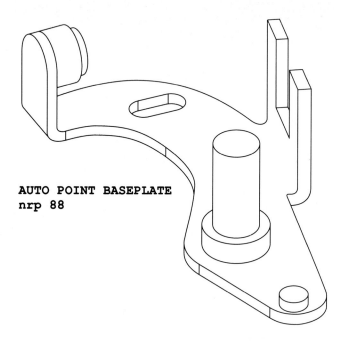

AUTO POINT BASEPLATE
nrp 88

3. Make a table and the other most appropriate graphic for illustrating each of the following. Draw the graphic and discuss the information contained in it.
   a. A comparison of the number of students obtaining jobs in their field compared to the total number of graduates.
   b. A comparison of the change in cost of a typical shopping list of foods over the past five years.
   c. A comparison of the number and percentage of students starting college and graduating in four, five, or six years.
   d. A comparison of six cereals on the basis of cost and nutrition.
4. Draw a bar graph with multiple bars and pie charts to illustrate the solid-waste information in the following table.

	Percent by Mass			
Component	City A	City B	City C	City D
Food waste	12	10	15	17
Paper	37	40	32	50
Cardboard	10	5	15	5
Plastic	3	6	5	4
Garden trimmings	10	12	14	9
Glass	8	9	8	7
Cans	6	7	7	6
Miscellaneous	14	11	4	2

5. Draw a line graph, bar graph, and pie chart to illustrate the size distribution of the sand in the table below.

% of Sand	Size (ft x 10^{-3})
1.10	3.28
6.6	2.29
15.8	1.77
18.6	1.51
17.6	1.25
19.0	1.05
15.0	0.88
4.0	0.77
2.3	0.59

6. Develop a histogram for the information in number 6.
7. Obtain information from a local industry and make an organizational chart showing job titles and the chain of command.
8. Obtain a spread-sheet software program that has graphic capabilities and use it to complete the graphics in number 5 or 6.
9. Draw a map of your state showing the number of students from each county who attend your school.
10. Draw a surface chart to illustrate the consumption of energy by the following forms of transportation for the past eight years.

39.5

	1971	1972	1973	1974	1975	1976	1977	1978
Cars	50%	50%	45%	40%	43%	45%	48%	50%
Trucks	19	20	20	18	17	19	23	25
Buses	10	12	17	20	19	16	14	10
Trains	16	15	15	17	15	15	10	10
Boats	5	3	3	5	6	5	5	5
Total Fuel	250	252	240	248	255	263	270	275

11. Draw a bar graph to illustrate the information in number 11.
    Divide each bar into the five transportation categories and
    make each bar represent the total amount of fuel used.

●40

DOCUMENTATION

KEY POINTS

- Document sources to prevent the charge of plagiarism.
- Take clear, careful notes when you research.
- Find out if your employer has a style sheet to guide you in documentation.
- Use the footnote system if you wish to add comments about a source.
- Use the numbered bibliography system for the briefest documentation.
- Use the author-date system to stress the date of the source's publication.
- Be sure to include the standard bibliographic material in each "Reference" entry, no matter which system of documentation you choose.

The term *documentation* has traditionally meant the giving of formal credit to sources, such as a professional paper or article, a book, or report, from which an author has obtained information and quotes.

Recently, the meaning of the term has broadened. In business and industry, the term *documentation* also refers to information that is recorded, usually on paper, and given to other people to read. Such documentation includes technical reports, proposals, specifications, presentations, journal articles, marketing research, and internal documents, such as memos and letters. As in the traditional use of the term, documentation gives formal credit to a source, in this case the writer, who can be praised or blamed as problems relating to the information arise. However, the concern here is with the narrower meaning—the citing of sources used by the author of a technical paper.

For several reasons it is necessary to give formal credit to sources of information and quotations that you use in your writing. First of all, the citation protects you, the writer, from the charge of plagiarism—stealing someone else's work or concepts and passing them off as your own. Many writers do not realize that paraphrasing (a restatement of a text or passage) may still be plagiarism. No matter how complete the rephrasing, if the information is garnered from some source and is not general knowledge, that source must be cited. Certainly, whenever the exact phraseology or even a striking word is borrowed, quotation marks must be employed and the source cited. Even if the information you are using is common knowledge to be found in any magazine, you must still give credit to your source if you copy the exact words or phrases.

What then, besides direct quotations, needs to be credited? Opinions, judgments, conclusions, ideas, and concepts that belong to a particular person need to be given formal credit. If disagreement exists about whether something is common knowledge, it needs to be credited. The findings and conclusions of particular experiments and case studies need to be credited. In general, document material where speculation is involved. If you are not sure whether to cite a source, play it safe and refer to it.

The second reason for giving formal credit to information sources is that it lends authority to your statements since the person whose work you are citing may be more of an expert on the subject than you are. Even if this is not true, the citation says that another person can corroborate your opinions. You are also telling the reader where to look for even more evidence of the validity of what you say.

Third, you provide the reader with sources for further research on a subject in which that person has already shown interest.

Depending on the system of notation you choose, two or three procedures are used to give formal credit to sources: reference within the text, footnote or endnote, and bibliographic material. So many documenting systems exist that many technical writers feel frustrated about selection. Most companies have their own style sheet that gives their preference in notation systems. For article or book publication, most publishers also provide style sheets to help in such matters. Academic departments or colleges within a university usually have style sheets that select one system or another.

Despite all the variations, three documentation systems are widely used. If you are not given a style sheet that spells out a particular system, you may want to adopt one of these. The Modern Language Association (MLA) has been very active for years in setting standards for research notation. The method that they recommended for years uses superscript numbers within the text that are keyed to notes placed either at the bottom of the page (footnotes) or at the end of the text (endnotes). The bibliography entries are put at the end of the main text or after the endnotes. See Figure 40-3, Sample Report One, at the end of this section for an example.

Recently the MLA has advocated a new system of notation that eliminates footnotes and endnotes. Replacing the superscript number within the text is parenthetical information: the author's name and the date of the source's publication as well as its page number. The bibliographic entries at the end of the text give the full information on the source. See Figure 40-4, Sample Report Two, at the end of this section.

The third popular notation system involves numbering the bibliographic entries and referring back to them by numbers inside parentheses within the text. Usually the page number where the information is found is also placed within the parentheses. See Figure 40-5, Sample Report Three, at the end of this section.

Many companies, schools, and other groups adopt one of these three styles, sometimes making their own particular modifications.

40.1   AVOID PLAGIARISM.

Give credit to the source from which you quote any word, phrase, sentence, or larger unit. Give credit to any source from which you take an opinion, judgment, conclusion, idea, or concept.

40.2   WRITE THE NAME OF YOUR SOURCE AND ALL THE BIBLIOGRAPHIC INFORMATION ABOUT IT ON A BIBLIOGRAPHY CARD.

Begin your research by gathering a list of good source materials on the topic. For each source you use, make out an index card with

FIGURE 40-1 **EXAMPLE OF A BIBLIOGRAPHY CARD**

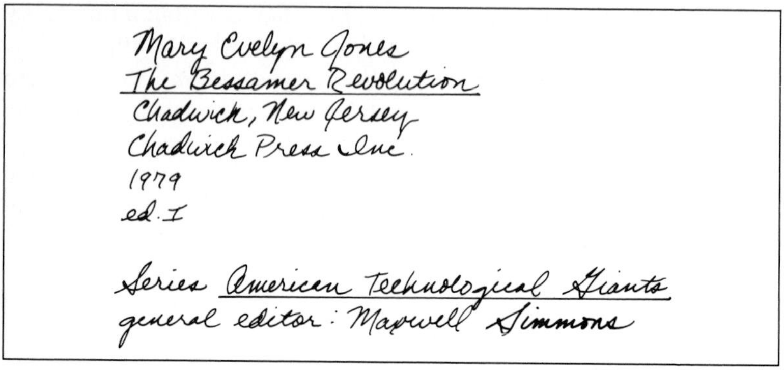

Mary Evelyn Jones
The Bessamer Revolution
Chadwick, New Jersey
Chadwick Press Inc.
1979
ed. I

Series American Technological Giants
general editor: Maxwell Simmons

all the bibliographic information necessary for your writing. Include the following:

1. Full name of author(s)
2. Title of book, article, or pamphlet
3. Title of publication the article appears in
4. Name of the editor, compiler, translator
5. Number of the edition or issue
6. Number of volumes if it is part of a set
7. Name of the series if it is part of a series
8. Place of publication
9. Name of publisher
10. Date of publication
11. Page numbers where article appears

Figure 40-1 is an example of a completed bibliography card.

40.3 TAKE CAREFUL NOTES, PUTTING QUOTATION MARKS AROUND COPIED WORDING AND NOTING CLOSE PARAPHRASING.

Much plagiarism is a result of poor note-taking procedures. The best method is to put quotation marks around quotes and to put other information entirely in your own words. Don't put more than one idea or concept on a single index card. Try to think ahead and consider how much information you would reference in any one place in your text. Restrict the information on the card accordingly.

FIGURE 40-2 EXAMPLE OF NOTE-TAKING ON AN INDEX CARD

Jones, Bessamer   p. 182

"Bessamer was a technological giant among giants," said Thomas A. Edison.

Edison also said he thought the bessamer process was somewhat inefficient, "more inefficient than necessary."

a. Put your notes from a source directly onto an index card.

The use of single cards for note-taking facilitates the arrangement of material later for incorporation into the text. If you have put only one concept or piece of information on a single card, then when you write the paper you can easily sort the cards in the order you will need them. You will be able to find specific research and to use all of it. The note card in Figure 40-2 demonstrates this type of note-taking.

40.4 CHOOSE AMONG THE THREE BASIC DOCUMENTATION SYSTEMS WITH CARE.

a. Use the footnote/endnote and bibliography system for thorough referencing.

In this documentation system, use a superscript number for the textual reference and key it to either a footnote or an endnote. Wherever the referenced note appears, if it is the first instance of reference to a particular source, provide the complete information on that source. (You give the same information later in the bibliographic entry, but in slightly different form.) If the source is referred to again, the footnote/endnote should list only the author and the page number where the information appears. If the cited author has two sources used in the paper, include the title of the work as well in later references.

This documentation system should be used if you wish to make comments of any sort in the footnotes/endnotes. In the footnote style, this system is also excellent if the cited source needs to be readily available to the reader, who has only to glance at the bottom of the page. For the same reason, footnotes are convenient for inserting comments.

EXAMPLE

A portion of a text follows:

> Due to the additives, other chemicals, and the composition of oil, used engine oil is now considered a hazardous waste product. In general, a substance is classified as hazardous waste if it is ignitable, reactive, corrosive, or toxic, "or if the EPA is out to get the company."[1]

---

[1] Stuart James, "Hazardous Waste," *Industry* (October 1982), p. 66. Mr. James is the corporate attorney for Ramon Oil Corporation.

b. Use the numbered citation and bibliography system for the briefest referencing.

In this documentation system, you number the bibliographic entries that are listed after the text as "References" or "Works Cited." When you use information from one of the sources listed, you put in parentheses the number assigned to that source. Although many people who use this system put only the source number in the parentheses, others put also the page or pages where the information is found. Do not use footnote/endnote entries in this system.

Because this system gives the reader the least information about the source within the text itself, you certainly have no room for comments on the reference.

EXAMPLE

A portion of a text follows:

> More specifically, Docket 80-90 was adopted in 1983. It was set into motion in 1984 and no new station licenses are expected to be granted until 1987 or later (3).

Works Cited

1. Ably, Ronald H. *Communications and the Policymakers.* New York: B.H. Adams Publishing Co., Inc., 1981.

2. Cunningham, Janet Day. "FCC Deregulation and the Lawyers." *Radio* 110 (October 1986), pp. 47–51.

3. Mazyruski, Bernard. *Can the Small Station Survive?* Cramdon, New Jersey: Baker and Baker, 1986.

4. Razymud, Nidal E. *You and Your Radio.* New York: Allendale Publishing Co., Inc., 1985.

c. Use the author-date citation and bibliographic system to present the most important information within the text.

The author-date system of documentation is now advocated by the MLA and is rapidly being adopted by writers in general. In this system the reference is given in parentheses within the text after the information that needs formal credit. The reference includes the author's last name, the year of publication, and the page numbers where the information can be found. If the source does not acknowledge an author, then provide a shortened version of the title of the work. Omit in the reference whatever information is given in the text.

One of the best features of this documentation system is that the date of the source's publication is given within the text. Technical and scientific communications must present the most current information in a rapidly changing world. Publication dates are, therefore, important in a citation.

However, a problem is created in this system because two numbers lie close together—the date and the page number. Writers use various means to solve this problem. Some separate the items with commas; others use a colon between the date and the page number. Occasionally writers use a comma and the abbreviation *p.* before the page number. In other cases, the system is basically adhered to, but the parenthetical information lacks one of the numbers.

The author's name or its necessary substitute is keyed to the bibliographic entry at the end of the text.

EXAMPLE

A portion of a text follows:

> Creel surveys have been utilized since the 1920s to evaluate the effects of fishery management. They have not always been well done (Carlin 1943: 57). The creel census is still regarded as a valuable tool of fishery management (Fine, *Survey*, 1977: 104). Most creel surveys are limited in application to individual bodies

40.5

of water (Fine, *Creel*, 1980: 32). A state survey would give valuable aid in decisions about hatchery production.

Works cited:

Carlin, John J. *Sampling Problems in Creel Census.* New York: Daedalus Publications, Inc., 1943.
Fine, Solomon P. "Creel Census Need." *Fish Management* 105 (Fall 1980), pp. 29–34.
———. "Creel Survey." *Fish Management* 96 (Summer 1977), pp. 100–106.

40.5 NO MATTER WHICH DOCUMENTATION STYLE YOU USE, THE BIBLIOGRAPHIC ENTRIES HAVE THE SAME TYPE OF INFORMATION.

The bibliographical entries vary in actual information depending on the kind of source, but they have the same type of information whatever documentation system is used. The numbered citation and bibliography system adds numbers before each entry—the only main difference among any of the systems. Also, the placement of the date varies. The following are standard bibliographical entries for various types of information sources, as a quick reference.

EXAMPLES

A book:

Hardy, Robert K. *Technology Today.* Phillipsburg, Maryland: National Publishing House, Inc., 1984.

A book issued by an organization:

Department of Justice. *Drug Use in the Sixties.* Washington, D.C.: U.S. Government Printing Office, Report A23397-81, 1972.

A book compiled by an editor:

Zebronski, Annette M., ed. *Essays on Norbert Wiener.* Mardge Hill, Colorado: Mardge Hill Press, 1967.

An article in a book:

Dean, Herbert and Robert Mays. "Electron Microscopes." *The Modern
    Laboratory.* Ed. Wilma Waid. St. Petersburg: Everglade Publishing
    House, Inc., 1980, 341–49.

A periodical article:

Raderman, Sandra U. "Business As Usual." *Enterprise* 41, 1983,
    42–45.

An anonymous article:

"Much Ado About Nothing: Formless Art." *Art World* 92, 1983, 56–61.

A newspaper article:

Morris, Sandamal. "The Nuclear Threat: Clear Skies Today." *Lakeland
    Times,* June 12, 1984, sec. B, 31.

A personal interview:

Smith, Rock Jones, President of Wave Electronics, Dallund, Wyoming.
    Interview with author. Dallund, August 27, 1986.

Note: One of the most popular variations in bibliographic entries is
that the publication date follows the author's name or the first entry
if it has no author. Also, the day can precede the month.

EXAMPLE

Jones, Rupert. 1985. *Optimum Bends in Optical Fibers.* Paris,
France: City Lights Publishing Co.
27 August 1989

●**EXERCISE 1**   Go to the library and copy five examples of bibliographic information from the author-subject catalogue. Write up the
information as a "Works Cited" list.

●**EXERCISE 2**   Find a book that has the footnote/endnote type of
documentation system. Go through one chapter, changing the docu-

mentation of five or ten entries to first the author/date system and then the numbered citation and bibliography system.

●**EXERCISE 3**   Write bibliographic entries for the following:

1. A book, titled *The Largest Cyclotron*, was published in 1983 and written by Shirley Markham. The Hyland Press of New York brought it out.
2. An article entitled "Furniture for the Office: 1985," written by Arnold Maier, was published by *Business Supplies Magazine* on March 14, 1984, issue 96.
3. Dr. Robert L. Stone, Chairperson for the Electrical Engineering Department at Dolliard State University, Dolliard, New Mexico, granted Al Bowdoin an interview at his office on September 18, 1986.
4. Mr. David Keeler, IBM sales manager for the Blanchard, Iowa, office, wrote an article on "The U.S. Future for IBM." It was published in *Economics Today* in the June 1982 issue.
5. Mr. Robert Dennison edited a book that reprinted articles on large corporations' future plans, *Big Business and the Future*, published by Reynolds and Sons out of Carmel, California, in 1985.

FIGURE 40-3   SAMPLE REPORT ONE, USING ENDNOTES AND BIBLIOGRAPHY

A STUDY OF MILK YIELD EXPECTATION
WITH ADMINISTRATION OF THE
BOVINE GROWTH HORMONE IN THE
HOLSTEIN HERD AT THE PIONEER STATE
UNIVERSITY FARM*

Introduction

Reference to a book by one author

Reference to a journal article

Second reference

Reference to a government publication

Administration of bovine growth hormone (bGH, bovine somatropin) to dairy cows substantially increases milk yields. (1) Somatropin, which is a naturally occurring pituitary hormone, enhances feed consumption and utilization. The resulting increase in glucose in the bloodstream and in increased blood flow to the mammary vein causes an increase in milk production. (2) In the past, research with bGH has involved natural somatropin derived from the anterior pituitary tissue of slaughtered animals. The available quantity of anterior pituitary tissue, however, was insufficient to allow a substantial increase in the nation's milk production. (3) Researchers recently discovered a synthetic method of producing recombinantly derived methionyl bovine somatropin (MBS). MBS can be produced in sufficient amounts to allow commercial use in dairy herds. Synthetic bGH, which may be commercially available by 1988, will be reasonably expensive and will result in increased feed consumption. (4)

This report will present research on the expected increase in milk yield resulting from studies on the use of bGH in published studies and

* Reprinted by permission of Kent Weigel, Platteville, Wisconsin.

in the Holstein herd at the Pioneer State University farm. Since bGH research in other breeds of dairy cattle is limited, this study will involve only the Holstein herd.

Pituitary vs. Recombinant Somatropin

Reference to a book by several authors

A published research study compares the effects of pituitary bovine somatropin (PBS) and recombinantly derived methionyl bovine somatropin in dairy cows. (5) High-producing Holstein cows were injected for 188 days with either 27 mg/day of recombinant somatropin or 27 mg/day of pituitary somatropin. The MBS increased milk yield 36.2% over controls, while the PBS group showed a 16.5% increase over controls.

The MBS and PBS groups both increased milk production to a similar peak initially, but the PBS group declined in production much more rapidly than the MBS group. The reason for this phenomenon, which led to the significant difference in milk yield between the two groups, is unknown. It is possible that the response to synthetic commercial bGH will be equal or greater than the response to pituitary bGH has been in previous years.

Method of Administration

Reference to a book by two authors

Research of published material showed significant differences in milk yield response to bGH depending on the method of administration. (6) Pituitary-derived somatropin (25 IU) increased milk yield 32.2% when injected daily, but only 8.5% when injected on alternate days.

The effect of a daily subcutaneous injection of bGH was slightly more than predicted. An injection of 12.5 IU/day should yield 50 to 60% of the 25 IU/day response. However, an alternate day injection of 23 IU (12.5 IU/day) yielded only 26% of the response of a daily injection. Response to a continuous subcutaneous infusion of 25 IU/day was similar to the daily injection response. Based on this study, extending the injection interval beyond 24 hours diminishes the milk yield response, so commercial bGH use will involve a single daily injection.

Dosage

Second reference

A previously published study regarding the effect of bGH dosage level on milk yield involved daily injections of 5, 10, 25, 50, and 100 IU of pituitary somatropin in Holstein cows. Milk production increases ranged from 7.1% to 31.8% over controls. (7)

Somatropin rapidly increases milk yield at small dosage levels, but the response diminishes as the dose becomes large. A single injection of about 50 IU/day appears to maximize milk yield while minimizing bGH waste.

Expected Growth Hormone Effect on Milk Production in the Pioneer Holstein Herd

Reference to an unpublished study

June 1986 production records from the Pioneer State University Holstein herd show a rolling herd average (RHA) of 17,772 lbs milk/cow/lactation and a daily average of 52 lbs/cow/day for the 60-cow herd. Adjustment of these figures based on expected bGH responses from previous studies results in substantial expected production increases. (8)

Daily milk production will vary depending on the season and the stage of lactation of the majority of the herd. Milk yield increases due to bGH administration vary from 10 to 40% depending on the herd and the individual cow.

Most increases range from 20 to 30%; 25% is the
expected value most often quoted in literature.
Milk yield increases are greater in late lactation,
when the cow is in more positive energy balance.
Administration of bGH also increased fat test
slightly, but not enough for a significant
correlation. Based on this study, the Pioneer State
farm can expect an increase in milk production of
at least 25% due to bGH administration.

Notes

1. Chambers, William. *The Bovine Growth
   Hormone* (Boston: Manchester Publications,
   Inc., 1984), p. 15.
2. Bouzelmann, Susan M. "The Effects of bGH."
   Animal Lives, LXXV (October 1985),
   pp. 16–24.
3. Bouzelmann, p. 27.
4. National Industrial Research Forum. *Research
   and Results,* Studies in Farm Economics, No. 85
   (New York: National Industrial Research
   Forum, 1985), p. 107–32.
5. Baez, Robert, Cinnco, Lewis A., and Harold R.
   Schneider. *Comparing Bovine Growth
   Hormones* (Philadelphia: Penn and Sons
   Publishing Company, 1984), pp. 111–56.
6. Hernandez, Philippe, and Helen S. Lowe.
   Administration of Growth Hormones in Cows
   (Sacramento: Palmer and Palmer, Inc., 1985),
   p. 54.
7. Hernandez, p. 201.
8. Czamanchi, Peter, Chair, Department of
   Agriculture, Pioneer State University, 1985.

References

Baez, Robert, Cinnco, Lewis A., and Harold R.
    Schneider. *Comparing Bovine Growth*

Hormones. Philadelphia: Penn and Sons
        Publishing Company, 1984.
Bouzelmann, Susan M. "The Effects of bGH."
        Animal Lives, LXXV (October 1985),
        pp. 15–32.
Chambers, William. *The Bovine Growth Hormone.*
        Boston: Manchester Publications, Inc., 1984.
Czamanchi, Peter, Chair. Department of
        Agriculture, Pioneer State University, 1985.
Hernandez, Philippe, and Helen S. Lowe.
        Administration of Growth Hormones in Cows.
        Sacramento: Palmer and Palmer, Inc., 1985.
National Industrial Research Forum. *Research
        and Results,* Studies in Farm Economics, No.
        85. (New York: National Industrial Research
        Forum), 1985.

**FIGURE 40-4   SAMPLE REPORT TWO, USING NUMBERED
CITATIONS AND BIBLIOGRAPHY**

A STUDY OF MILK YIELD EXPECTATION
WITH ADMINISTRATION OF THE
BOVINE GROWTH HORMONE IN THE
HOLSTEIN HERD AT THE PIONEER STATE
UNIVERSITY FARM

Introduction

Reference to a
book by one
author

Administration of bovine growth hormone
(bGH, bovine somatropin) to dairy cows
substantially increases milk yields (3:15).
Somatropin, which is a naturally occurring
pituitary hormone, enhances feed consumption
and utilization. The resulting increase in glucose
in the bloodstream and in increased blood flow to

Reference to a
journal article

the mammary vein causes an increase in milk
production (2:16–24). In the past, research with

Second
reference

bGH has involved natural somatropin derived from the anterior pituitary tissue of slaughtered animals. The available quantity of anterior pituitary tissue, however, was insufficient to allow a substantial increase in the nation's milk production (2:27). Researchers recently discovered a synthetic method of producing recombinantly derived methionyl bovine somatropin (MBS). MBS can be produced in sufficient amounts to allow commercial use in dairy herds. Synthetic bGH, which may be commercially available by 1988, will be reasonably expensive and will result in increased feed consumption (6:107–32).

Reference to a
government
publication

This report will present research on the expected increase in milk yield resulting from studies on the use of bGH in published studies and in the Holstein herd at the Pioneer State University farm (1:111–56). Since bGH research in other breeds of dairy cattle is limited, this study will involve only the Holstein herd.

Pituitary vs. Recombinant Somatropin

Reference to a
book by several
authors

A published research study compares the effects of pituitary bovine somatropin (PBS) and recombinantly derived methionyl bovine somatropin in dairy cows (1:111–56). High-producing Holstein cows were injected for 188 days with either 27 mg/day of recombinant somatropin or 27 mg/day of pituitary somatropin. The MBS increased milk yield 36.2% over controls, while the PBS group showed a 16.5% increase over controls.

The MBS and PBS groups both increased milk production to a similar peak initially, but the PBS group declined in production much more rapidly than the MBS group. The reason for this phenomenon, which led to the significant difference in milk yield between the two groups, is unknown. It is possible that the response to synthetic commercial bGH will be equal or greater than the response to pituitary bGH has been in previous years.

Method of Administration

Reference to a book by two authors

Research of published material showed significant differences in milk yield response to bGH depending on the method of administration (5:54). Pituitary-derived somatropin (25 IU) increased milk yield by 32.2% when injected daily, but only 8.5% when injected on alternate days.

The effect of a daily subcutaneous injection of bGH was slightly more than predicted. An injection of 12.5 IU/day should yield 50 to 60% of the 25 IU/day response. However, an alternate day injection of 23 IU (12.5 IU/day) yielded only 26% of the response of a daily injection. Response to a continuous subcutaneous infusion of 25 IU/day was similar to the daily injection response. Based on this study, extending the injection interval beyond 24 hours diminishes the milk yield response, so commercial bGH use will involve a single daily injection.

Dosage

Second reference

A previously published study regarding the effect of bGH dosage level on milk yield involved daily injections of 5, 10, 25, 50, and 100 IU of pituitary somatropin in Holstein cows. Milk production increases ranged from 7.1% to 31.8% over controls (5:201).

Somatropin rapidly increases milk yield at small dosage levels, but the response diminishes as the dose becomes large. A single injection of about 50 IU/day appears to maximize milk yield while minimizing bGH waste.

Expected Growth Hormone Effect on Milk Production in the Pioneer Holstein Herd

June 1986 production records from the Pioneer State University Holstein herd show a rolling herd average (RHA) of 17,772 lbs milk/cow/lactation and a daily average of 52 lbs/cow/day for the 60-cow herd. Adjustment of these figures based on expected bGH responses from previous studies results in substantial expected production increases (4).

Reference to an unpublished study

Daily milk production will vary depending on the season and the stage of lactation of the majority of the herd. Milk yield increases due to bGH administration vary from 10 to 40% depending on the herd and the individual cow. Most increases range from 20 to 30%; 25% is the expected value most often quoted in literature. Milk yield increases are greater in late lactation, when the cow is in more positive energy balance. Administration of bGH also increased fat test slightly, but not enough for a significant correlation. Based on this study, the Pioneer State farm can expect an increase in milk production of at least 25% due to bGH administration.

Works Cited

1. Baez, Robert, Cinnco, Lewis A., and Harold R. Schneider. *Comparing Bovine Growth Hormones*. Philadelphia: Penn and Sons Publishing Company, 1984.
2. Bouzelmann, Susan M. "The Effects of bGH." *Animal Lives,* LXXV (October 1985), 15–32.
3. Chambers, William. *The Bovine Growth Hormone*. Boston: Manchester Publications, Inc., 1984.
4. Czamanchi, Peter, Chair. Department of Agriculture, Pioneer State University, 1985.
5. Hernandez, Philippe, and Helen S. Lowe.

Administration of Growth Hormones in Cows.
Sacramento: Palmer and Palmer, Inc., 1985.
6. National Industrial Research Forum. *Research
    and Results,* Studies in Farm Economics, No.
    85. (New York: National Industrial Research
    Forum), 1985.

**FIGURE 40-5   SAMPLE REPORT THREE, USING AUTHOR/DATE
CITATIONS AND BIBLIOGRAPHY**

A STUDY OF MILK YIELD EXPECTATION
WITH ADMINISTRATION OF THE
BOVINE GROWTH HORMONE IN THE
HOLSTEIN HERD AT THE PIONEER STATE
UNIVERSITY FARM

Introduction

Reference to a
book by one
author

Reference to a
journal article

Second
reference

Administration of bovine growth hormone
(bGH, bovine somatropin) to dairy cows
substantially increases milk yields (Chambers
1984:15). Somatropin, which is a naturally
occurring pituitary hormone, enhances feed
consumption and utilization. The resulting
increase in glucose in the bloodstream and in
increased blood flow to the mammary vein causes
an increase in milk production (Bouzemann
1985:16–24). In the past, research with bGH has
involved natural somatropin derived from the
anterior pituitary tissue of slaughtered animals.
The available quantity of anterior pituitary tissue,
however, was insufficient to allow a substantial
increase in the nation's milk production
(Bouzemann 1985:27). Researchers recently
discovered a synthetic method of producing
recombinantly derived methionyl bovine
somatropin (MBS). MBS can be produced in
sufficient amounts to allow commercial use in

dairy herds. Synthetic bGH, which may be commercially available by 1988, will be reasonably expensive and will result in increased feed consumption (National Research 1985: 107–32).

Reference to a government publication

This report will present research on the expected increase in milk yield resulting from studies on the use of bGH in published studies and in the Holstein herd at the Pioneer State University farm. Since bGH research in other breeds of dairy cattle is limited, this study will involve only the Holstein herd.

Pituitary vs. Recombinant Somatropin

A published research study compares the effects of pituitary bovine somatropin (PBS) and recombinantly derived methionyl bovine somatropin in dairy cows (Baez 1984:111–56). High-producing Holstein cows were injected for 188 days with either 27 mg/day of recombinant somatropin or 27 mg/day of pituitary somatropin. The MBS increased milk yield 36.2% over controls, while the PBS group showed a 16.5% increase over controls.

Reference to a book by several authors

The MBS and PBS groups both increased milk production to a similar peak initially, but the PBS group declined in production much more rapidly than the MBS group. The reason for this phenomenon, which led to the significant difference in milk yield between the two groups, is unknown. It is possible that the response to synthetic commercial bGH will be equal or greater than the response to pituitary bGH has been in previous years.

Method of Administration

Research of published material showed significant differences in milk yield response to

Reference to a
book by two
authors

bGH depending on the method of administration
(Hernandez 1985:54). Pituitary-derived
somatropin (25 IU) increased milk yield by 32.2%
when injected daily, but only 8.5% when injected
on alternate days.

The effect of a daily subcutaneous injection of
bGH was slightly more than predicted. An
injection of 12.5 IU/day should yield 50 to 60% of
the 25 IU/day response. However, an alternate day
injection of 23 IU (12.5 IU/day) yielded only 26%
of the response of a daily injection. Response to a
continuous subcutaneous infusion of 25 IU/day
was similar to the daily injection response. Based
on this study, extending the injection interval
beyond 24 hours diminishes the milk yield
response, so commercial bGH use will involve a
single daily injection.

Dosage

A previously published study regarding the
effect of bGH dosage level on milk yield involved
daily injections of 5, 10, 25, 50, and 100 IU of
pituitary somatropin in Holstein cows. Milk
production increases ranged from 7.1% to 31.8%
over controls (Hernandez 1985:201).

Second
reference

Somatropin rapidly increases milk yield at
small dosage levels, but the response diminishes
as the dose becomes large. A single injection of
about 50 IU/day appears to maximize milk yield
while minimizing bGH waste.

**Expected Growth Hormone Effect on Milk
Production in the Pioneer Holstein Herd**

June 1986 production records from the
Pioneer State University Holstein herd show a
rolling herd average (RHA) of 17,772 lbs milk/
cow/lactation and a daily average of 52 lbs/cow/

40.5

Reference
to an
unpublished
study

day for the 60-cow herd. Adjustment of these figures based on expected bGH responses from previous studies results in substantial expected production increases (Czamanchi).

Daily milk production will vary depending on the season and the stage of lactation of the majority of the herd. Milk yield increases due to bGH administration vary from 10 to 40% depending on the herd and the individual cow. Most increases range from 20 to 30%; 25% is the expected value most often quoted in literature. Milk yield increases are greater in late lactation, when the cow is in more positive energy balance. Administration of bGH also increased fat test slightly, but not enough for a significant correlation. Based on this study, the Pioneer State farm can expect an increase in milk production of at least 25% due to bGH administration.

Works Cited

Baez, Robert, Cinnco, Lewis A., and Harold R. Schneider. *Comparing Bovine Growth Hormones.* Philadelphia: Penn and Sons Publishing Company, 1984.

Bouzelmann, Susan M. "The Effects of bGH." *Animal Lives,* LXXV (October 1985), 15–32.

Chambers, William. *The Bovine Growth Hormone.* Boston: Manchester Publications, Inc., 1984.

Czamanchi, Peter, Chair. Department of Agriculture, Pioneer State University, 1985.

Hernandez, Philippe, and Helen S. Lowe. *Administration of Growth Hormones in Cows.* Sacramento: Palmer and Palmer, Inc., 1985.

National Industrial Research Forum. *Research and Results,* Studies in Farm Economics, No. 85. New York: National Industrial Research Forum, 1985.

●41

DEFINITION

KEY POINTS

- Select the kind and length of definition used according to the needs of the audience.
- Choose extended definition if the reader needs a full grasp of a term's meaning.
- Write sentence definitions with care not to place the term in too broad a class, or to repeat the term or any derivative of it in the definition.
- Use a variety of methods to define a term.
- Choose carefully the placement of the definition.

The ability to define terms clearly and to know what to define in a given situation is, for the writer, a fundamental skill. This is true for no one more than for the technical writer, whose main goals are clarity and accuracy.

People in all professional, business, and industrial fields need definitions and must often write them. First of all, someone defined your job and the duties it entails. Second, the company has manuals that define your benefits and your obligations. You may need to define company policies and procedures for government officials or clients. Professionals frequently write articles for publication in journals; most of these require at least some definition. Definitions enter into every kind of technical writing: memos, letters, manuals, proposals, informal and formal reports, and technical articles and papers.

The use you make of definition depends largely on the audience being addressed. Every field of endeavor has a specialized vocabulary meaningful only to the people in that field. Terms that need not be defined for them would need to be defined for others. In general, a nontechnical audience needs at least some terms defined in any technical communication. It should be clear, however, that a nontechnical audience can absorb only a few of such definitions. The semitechnical reader usually knows enough about the technical subject to handle quite a few definitions, perhaps even a large glossary that precedes the text. Some audiences include people with different levels of knowledge of a technical subject's vocabulary. In this case, you must make definition choices that accommodate the least technical member of the audience, or write various sections of the text for different audience levels.

Other reasons for definition include the need to stipulate, to say which of several possible meanings is meant in this particular writing, or to acknowledge that a term may have recently undergone a change in meaning.

The three basic types of definition are the *parenthetical*, the *sentence*, and the *extended*. *Parenthetical* definition gives a synonym or phrase definition "in passing." The *sentence* or formal definition resembles dictionary entries in style and content. Such definitions may be placed in the text itself, in a glossary, as mentioned above, in footnotes or endnotes, or in an appendix. Because an audience may not take the trouble to read definitions that are not incorporated into the text, the writer usually considers it necessary to give at least a brief definition. An *extended* definition may require one or several paragraphs of a text or be the focus of the entire text, usually defining a number of approaches.

41.1 LET THE AUDIENCE'S NEED DETERMINE THE AMOUNT AND KIND OF DEFINITION.

a. Define new terms.

The most obvious use of definition is to explain a term new to the audience. The audience may be nontechnical, semitechnical, or highly technical. They may be job trainees, students, or employed professionals. The term that they need to understand may be a standard term for the field or a new term to everyone. In these cases, extended definition is necessary, whether this means one, two, or three paragraphs, or a whole paper. You can employ one or several methods.

EXAMPLE

Optical Communication Systems

A light-emitting diode (LED) is an electrical device distinguished by low cost, voltage, and power requirements. It is used in various ways to signal information and is an important part of many optical communication systems. Light from the LED is directed into fibers, which replace the wires of the old communication system. The light travels through the fiber like water through a garden hose. This light contains information in the form of code. Advantages of the new optical communication system include economy and signal quality.

b. Explain technical terms for nontechnical audiences.

One of your primary tasks is to be able to make technical material accessible to the least technical readers. The vocabulary needed is described as somewhere around the fifth- to the eighth-grade level. In other words, most people's everyday vocabulary is largely in place by the time they become teenagers. Writers who wish to explain complicated technical matters to the uninitiated must find a common-ground vocabulary. Certainly no symbols or uncommon abbreviations can be used.

EXAMPLES

A numeric-control-tool programmer is a computer specialist who plans the steps in a computer program that controls an automatic machine's cutting path during the making of a product. The programmer is concerned with how accurately the machine—laser, lathe, drill, robot, etc.—is cutting out parts.

Elevators have been around a long time. The invention in the 1800s of the hydraulic system of water pressure enabled engineers to make practical elevators—hydraulic elevators. These were operated by means of a vertical plunger that traveled in a cylinder that extended into the ground to a depth exactly equivalent to the distance the elevator traveled. The plunger, with the elevator car attached to its top end, moved up and down by means of the changes in water pressure governing it.

c. Define only highly technical terms for the semitechnical reader.

The semitechnical reader may be a person who has much practical experience with technical matters or a student who is preparing to enter the field. The semitechnical reader usually understands basic concepts and terms, and can usually understand technical jargon to a degree, but needs high-level technical terms defined.

EXAMPLE

Superconductivity allows for unprecedented sensitivity in measurement. The superconducting ring fluxmeter (SRF) measures changes in magnetic fields ten orders of magnitude smaller than the earth's field. The SRF is a member of a larger class of devices called Superconducting Quantum Interference Devices, with interference meaning the interference of two or more quantum-mechanical-wave functions, each characterizing the situation in a separate piece of superconducting metal.

d. In highly technical writing define only the field's newest terms.

Technical communication among experts necessitates little definition, unless a new term has entered the field.

EXAMPLE

All of these drivers are called with the same interface: the row and column coordinates of the upper left corner of the form to be drawn and a pointer to a data structure describing the form are passed in registers BX, CX, and SI. The row coordinate is in the range 0 to 199, with limitations for specific drivers.

41.2 CHOOSE THE LENGTH OF DEFINITION BASED ON WHAT THE SITUATION REQUIRES.

a. Use a parenthetical definition when the audience understands the term but might not recognize it.

Use parenthetical definition when you believe that the term, not the meaning or concept of the term, is not clear. Usually parenthetical definition gives a familiar term for the unfamiliar one used in the text. Sometimes the parenthetical definition is a phrase rather than a substitute term. Although these short definitions are usually called parenthetical, they may be enclosed simply in commas.

EXAMPLES

> Rhexia (deer grass) grows only in wet ground.
> This remnant of the temple of Solomon (called the Wailing Wall) is revered by Jews as a place of pilgrimage.
> The ceremonial headdress (or War Bonnet) is used by some Plains Indians.
> Mealy-mouthed people, i.e., those unwilling to state their opinions directly, cause many of the world's woes.
> The proboscis, or prominent nose, characterizes many species.

b. Use a sentence or formal definition to give the reader a precise and essentially complete definition of a term.

The formal definition gives the reader a concise overview of a term's referential place and meaning. It has three parts: term, class, and differentia. The formal definition identifies the subject as belonging to a particular class of things and then tells how it differs from the other items in its class. Most dictionaries extensively employ this type of definition.

EXAMPLES

Term	Class	Differentia
A lathe is	a shaping machine	that spins a piece on a horizontal axis, cutting and bending it to the desired size.
A Bofors gun is	an antiaircraft weapon	that is double-barreled and automatic; it was first made in Bofors, Sweden.
An endomorph is	a mineral	that is found as an inclusion in another, as rutile is found in quartz.

Sentence definitions follow several rules that promote clarity:

1. Place the term in a specific class.
2. Sufficiently differentiate the term from other members of its class.

3. Do not use the term or any derivation of it in the class or differentia.
4. Do not use jargon or highly technical words.

In relation to rule 1, place the term in a specific class to increase understanding. If the term *psychiatrist* is put in the class of "person," the reader has learned very little. On the other hand, if you give the class as "a licensed physician," the reader learns more.

Rule 2 states that the differentia should be clearly distinguished. In the above example of *psychiatrist*, the differentia must distinguish the psychiatrist, a physician of the mind, from physicians of the body, such as medical doctors, osteopaths, and chiropractors.

Rule 3 reminds the writer that use of the term or some derivative of it within the definition defeats the sentence's purpose. If a digital watch is defined as a class of watches with the differentiation that it is digital, the reader learns nothing. The digital watch is a portable timepiece, usually worn on the wrist, that uses changing numerals for its display, rather than the moving hands that characterize analog displays.

If the reader is highly technical, rule 4 means less. For most readers, however, a definition full of jargon is not helpful. It does little good to learn that a *field* is the gradient of the potential if we do not know what either a gradient or a potential is.

c. Use an extended definition to ensure that the reader more completely understands a term.

Extended definition is used when your intent is to give your audience a thorough understanding of a term. You can employ any of the following methods: physical description, analysis of parts, and operating principle; examples; analogies and contrasts; history or background and the etymology (origin from other and/or earlier languages); antonyms and negative statements of what it is not; cause and effect discussion; illustrations.

EXAMPLES

Physical description, analysis of parts, and operating principle:

A pipe wrench is a hand tool for exerting a twisting force or torque, usually on a cylindrical surface. It is adjustable to work on objects of different sizes.

The body of the pipe wrench is made of forged metal from a mid-carbon steel and ranges from 6 to 18 or more inches long and 1/2-inch thick. The corners are rounded and smooth; the handle has a groove to make it lighter. The jaw is made of hardened steel and has 8 teeth cut to a 45-degree angle. The heel, also hardened steel, has 8 teeth cut at an angle. The direction of the teeth on the heel opposes the direction of the teeth on the jaw for better grip. The heel has several threads that a round thumb-screw screws onto to provide quick adjustment.

A pipe wrench works by applying a force to the end of the body (or handle) in the direction of the opening of the jaws. The body of the pipe wrench acts as a lever arm to increase the amount of torque delivered to the object. The teeth on the heel and jaw grip the round surface. The wobbly jaw is designed to cause the grip to tighten as force is applied to the handle. The force applied at the end of the handle is transmitted to the heel and jaws, which act on the round surface to turn the object.

A drawing to scale of the prior description would help the reader understand the definition and locate the parts.

Illustrations:

Show a plumber making a house call to work on plumbing with a pipe wrench. Portray an automobile mechanic using a pipe wrench on a tailpipe or other car part.

Comparisons and negations:

A pipe wrench is similar to many of the basic wrenches available that exert a twisting force. A pipe wrench has nothing to do with the repair or adjustment of smoking apparatus. A crescent wrench is also an adjustable hand tool for applying twisting force. The difference between a crescent and pipe wrench is that a crescent wrench is limited to use on flat surfaces, such as nuts or the heads of bolts that have four to eight sides. A pipe wrench can be used on flat surfaces but is designed specifically for round surfaces or nuts and bolts that have been rounded off.

Box and end wrenches are similar to a crescent, but they are not adjustable and can be used only on nuts and bolts that are the proper size. Other wrenches available, such as spanning wrenches, oil-filter wrenches, and torque wrenches, are designed for specific purposes. A pipe wrench is an all-purpose tool that has a broad range of applications.

History, background, and etymology:

> The word *wrench* comes from an Old English word *wrencan,* which means "to twist." The British call the tool a *spanner,* which emphasizes its stretching ability. The open-end adjustable wrench appears to be older than the closed wrenches designed for particular purposes. In their earliest form, open-end wrenches had either straight, angled, or S-shaped handles, were of wrought or cast iron, and first appeared around 1800. By the early 1900s sliding-jaw-type wrenches, such as the pipe wrench, were most popular, and they continue in popularity today.

Cause and effect discussion:

> Because of the teeth and the adjustable size of the jaw space, the pipe wrench can engage round rods and pipes, a much needed function.

41.3 CHOOSE CAREFULLY THE PLACEMENT OF DEFINITIONS.

Put definitions in one of four places in a text: the text itself, a glossary, footnotes or endnotes, or an appendix. Their placement reflects on your conception of their integral importance, and to some degree, on your perception of the audience.

a. Incorporate definitions into the text if the information is vital to the audience's understanding of the topic.

EXAMPLES

> This ruling applies only to investments made by you or relatives (members of the immediate family: mother, father, children, spouse).
>
> Civil engineers are involved in the design and construction of static structures. Static structures are nonmoving, like buildings, bridges, and dams. An engine is a moving structure.

b. Put definitions in a glossary if your audience is semitechnical and needs only to check definitions of technical terms to ensure accuracy.

Many reports include glossary sections before the full discussion of the methods, data, results, conclusions, and recommendations.

The reader of this detailed section is often a semitechnical person who can read the material with comprehension, but needs to check a definition now and then. The following glossary is part of a manual that was written for students enrolled in a beginning computer-science course. Note that all of the terms have a formal sentence definition.

DEFINITIONS

COMMAND A set of instructions or an instruction used to perform a specific duty. See appendix for editor commands.

CURSOR A blinking square on the screen that indicates where the next character typed in would be located.

EDIT A process of making corrections or changes to a file.

FILE A program or set of data that is saved in memory. Also a command to save a set of instructions or set of data in memory.

FILENAME A set of eight unique characters or numbers used to distinguish files from one another.

FORMAT A process designed to initialize or clear out a space on the mini-desk within the computer. This is only done the very first time on the computer or previous work will be erased.

HARD COPY A printed copy of a file or the printed output of a program.

LIBRARY Storage areas in the computer in which different files are stored. Your mini-disk is a library; there is also a help library and a different library for every language in the computer.

LOG ON/LOG OFF The process used to begin or end a session on the computer.

OUTPUT The result of executing a program that will produce a printed result of a display on the terminal's screen.

PASSWORD A unique set of four characters that is part of the user's identification on the computer.

PROGRAM A set of instructions joined together to perform a specific task.

SAVE A process used to move the file the user was working on into storage on the user's minidisk. *Save* is the same as the file command.

USER ID A unique combination of eight characters and numbers that is part of the user's identification on the computer.

c. Use footnote/endnote definitions when the material is of interest but not absolutely necessary.

Definitions placed at the bottom of the page in footnotes are more accessible than definitions placed at the end of the text. Endnotes are like footnotes, only all the notes are put in one place at the end of the text. Endnote definitions imply that the material is even less crucial than that placed in footnotes. Neither placement gives the idea the definitions are of necessity, at least to the majority of readers.

d. Place definitions in the appendix for the select reader.

All material in the appendix is for readers who have a special need to investigate the subject more thoroughly than most of the readers of the text. Of the four placements, appendix definitions imply the least amount of necessity for the reader.

●**EXERCISE 1** Write an extended definition of the professional field you hope to enter or are already in.

●**EXERCISE 2** Find a semitechnical writing and rewrite it for nontechnical readers, interpolating the necessary definitions.

●**EXERCISE 3** Find a magazine or journal in your field, then find an article written for experts in the field. Rewrite it for a friend who is not in your field, incorporating the necessary definitions.

●**EXERCISE 4** Write one-sentence formal definitions of the following: byte, word processor, electron, personnel manager, corporation, joint ownership, asbestos, binoculars, and solar eclipse.

●42

DESCRIPTION

- Use technical descriptions as aids to other types of writing.
- Know the subject thoroughly before you attempt to write a technical description.
- Use the introduction to give the reader an overview and to orient the reader to the subject.
- Choose spatial, functional, or assembly order carefully.
- Make the conclusion brief.
- Use graphics in general to enhance text, not to replace it.

In general, a technical description gives physical characteristics of an object or mechanism. A mechanism is a device that has several identifiable parts working together as a system. Although a single object is the usual subject of a technical description, other possible subjects include substances (like paint), systems (like ignition systems), and locations (like the site for a new plant).

Graphic aids often accompany the verbal description. A technical description rarely appears as a separate writing. Most of it is used to describe pieces of equipment and is incorporated into manuals, proposals, and informal and formal reports. Technical descriptions are used to give operating and maintenance instructions in manuals; to describe needed equipment in proposals; to describe the set-ups, locations, and involved equipment of experiments for informal and formal reports. Sales brochures make ample use of technical descriptions to show customers potential purchases.

As in all aspects of technical writing, you must determine how much the readers know about the subject. For example, does the reader know what the subject looks like but less about how it works? Does the reader need sufficient description to build or operate the mechanism? Is the reader going to repair and maintain the mechanism? All such audience considerations are very important in technical descriptions.

42.1 GIVE ALL THE PERTINENT INFORMATION ABOUT THE PHYSICAL CHARACTERISTICS OF A SUBJECT.

a. Determine the properties that you must describe.

The main purpose of technical description demands that you cover a subject's major physical features. To do so, answer all of the following questions about the subject:

1. What is it? How is it defined?
2. What is its purpose or general function?
3. What does it look like? What is its size, shape, color, weight, height, age?
4. What are its characteristics? What is its flammability, density, expected life, method of production?
5. How does it work? What is the operating principle or theory? How is it assembled or taken apart? What is the operating cycle?
6. What are the major divisions or principle parts?

b. Acquaint yourself thoroughly with the subject before you write any part of the description.

First of all, be sure that you are thoroughly acquainted with the particular subject at hand. If you do not know what material it is made of, you must find out. If you do not know its durability, you should find out. Perhaps its function and operating principles are not immediately obvious. It helps if you can disassemble the object, but sometimes you must rely on interviews with designers and technicians to find this out. You may not be acquainted with its range of functions or know for sure what other mechanisms are available that

42.2

resemble this one but are slightly different. You need to know this to understand whether to offer a generic description or a particular one. You also need to know the class of objects the mechanism is in, in order to define it.

42.2   PLACE SUFFICIENT DESCRIPTION IN THE INTRODUCTION TO GIVE AN OVERVIEW OF THE SUBJECT.

Whether the technical description is a separate writing or, as is usual, part of a longer writing, it will have an introduction. The introduction must accomplish its purpose of orienting the reader about the object so that the more detailed description to follow is more understandable.

a. Let the writing situation govern the length and format of the introduction.

If your description is part of a larger writing, you may have restricted room to accomplish the description. The other consideration is audience; the amount of overview necessary may well be governed by the need for such information. Usually write the introduction in nontechnical terms, assuming that a more technical person needs little or no introduction.

b. Give certain essential information in the introduction, modifying the discussion to suit the writing situation.

Introductions to technical descriptions usually include the following:

1. A formal (sentence) definition of the subject
2. A statement of the purpose, use, or function of the subject
3. A very general description of the subject intended to give the reader a visual image of it
4. A statement about the principle or theory of operation
5. A list of the major parts, divisions, or assemblies of the subject

You can modify these five categories to fit your writing situation. Write several paragraphs that merge some of them. Or they may each receive a subheading under the general heading of "Introduction." The sequence may follow the given list or change to suit the situation.

42.3 GIVE DETAILED DESCRIPTIONS OF THE MAJOR PARTS AND ASSEMBLIES IN THE BODY OF THE TECHNICAL DESCRIPTION.

After you end the introduction with a list of the major parts, you are ready to discuss each of these parts in sufficient detail so that, say, a technician could build the part if necessary. In fact, you usually address the body of your description to a semitechnical reader who wants or needs the amount of detail it offers. Semitechnical readers would include those using manuals to maintain, repair, or operate a mechanism, as well as technicians who might produce it. The audience also may include a person who must assemble the mechanism but who does not always need the amount of detail useful to make or replace parts.

a. Consider carefully the order in which to describe the major parts and assemblies.

The first consideration is the order in which to describe the parts. It must be the same order as the introduction list of major parts.

The three basic ways to arrange the body of the description are spatial, functional, and assembly.

If you order the parts according to the subject's *spatial* appearance, you choose to order them as a viewer's eye might see it, starting with, for example, the top, moving to the body, and then the bottom, or starting with the left side, then going on to the middle, then ending with the right side. Whatever order you decide is the natural one a viewer would use when first seeing the object—you must present the parts in an orderly sequence, moving from top to bottom, bottom to top, one side to another side. A C-clamp would be best described spatially.

The second possible arrangement accords with the *function* or the order in which the parts would be engaged in an operating cycle. This is helpful when the look of the mechanism is not as important to the reader's comprehension of the object as is the function. For example, a fishing reel would best be described according to the way the parts are used in operating it.

The third sequence is in accord with the way the mechanism would be *assembled*. This is effective when many of the parts are hidden and can be revealed only when the part is disassembled. A cartridge pen would be more understandable done in assembly order.

b. Describe each major part thoroughly in the body of the description.

Each major part should have the following in its description:

1. A formal (sentence) definition that tells what the part is and its purpose. This also may be a list of subparts if there are any.
2. A detailed description of its appearance. This includes the size, shape, color, material, finish, weight, and age. Then describe physical characteristics, such as flammability, density, durability, and method of production.
3. Its location and functional relationship to other parts, especially to the part to be described next.

Include these same details for each subpart.

Note: If the mechanism is made entirely of one material, mention this in the introduction and omit making individual references to material in each major part.

42.4   MAKE THE CONCLUSION OF THE DESCRIPTION BRIEF.

The conclusion for a technical description is never long, but it serves a worthwhile function. After reading through the detailed descriptions of each major part, the reader needs to see the mechanism as a whole again. This is usually accomplished by giving assembly procedures and then telling one complete operating cycle for the mechanism. Write the conclusion with little or no passive voice (see Section 26).

42.5   USE GRAPHICS TO ENHANCE THE VERBAL DESCRIPTION.

The number of graphics that accompany the verbal text of a technical description is, to a large degree, a matter of personal preference and, in some cases, a matter of personal skills. Graphics should be used to help the reader become oriented with the mechanism so that the details are meaningful. Therefore, include a visual presentation of the mechanism in the introduction. You may profitably use graphic depictions of each of the major parts, perhaps drawing the part to scale and giving dimensions on the drawing. Exploded-view, cut-away drawing, diagrams, and other types of figures can be very helpful, especially when a part is not visible in the overview graphic depiction. Graphics most help the reader's visualization and understanding when used to show certain proportions and physical relationships, dynamics and simultaneity, and hard-to-see or "invisible" parts (such as electric current), all of which can create a clutter of detail if presented exclusively in paragraph form. But do not let the graphic aids do your verbal work for you. Render the description in words, letting the graphics clarify your statements. Figure 42-1, beginning on the next page, is an example of a technical description.

42.5

FIGURE 42-1 EXAMPLE OF A TECHNICAL DESCRIPTION

TECHNICAL DESCRIPTION OF A PLUMB BOB

Introduction

A plumb bob is a weight that is hung at the end of a line or string suspended from a height in order to determine verticality and for reference in angle measurement. The plumb bob operates on the relatively simple gravitational principle of weight. As it is hung vertically by a string, the earth's gravitational force acts on the weight to bring the string into an absolutely vertical line. The weight of the bob keeps it stationary over a specific location on the ground. Verticality of an edge or surface is determined by visually comparing it to the plumb-bob line.

The plumb bob is made up of four major parts: string, threaded cap, bob, and threaded tip.

Description of Major Parts

String

The string supports the weight of the plumb bob and provides a vertical line of sight to the surveyor. It is made of durable nylon

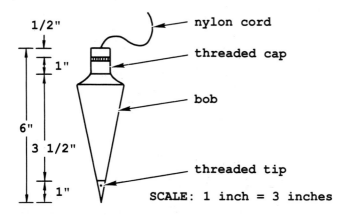

Figure 1-1: Front View of Plumb Bob

fiber that will stretch, and is manufactured in diameters of .0313 to .25 inches. A common color is bright orange, as the string must be visible from a distance. Directly below the string is the threaded cap. The string is inserted into this cap and thus attached to the plumb bob.

Threaded Cap

The threaded cap is a connection piece that provides a link between the string and the bob. It is cylindrical with a diameter of .5 inches, and length of 1.5 inches. A .25-inch vertical hole is drilled through the cap where the string is to be threaded and knotted securely. The cap is made of brass and has a smooth, machined surface. A raised, knurled brass ring .5 inches from the top provides a grip when the cap is being screwed. The cap is threaded at its bottom, allowing it to be screwed into the bob.

Bob

The bob provides the plumb bob with the gravitational weight needed to remain steady over a point. The bob is made of brass and is manufactured in weights of 10 to 32 ounces. It has a smooth, machined surface that, when polished, assists in making it visible in dark areas. It is cone-shaped, with a decreasing diameter from the top for a length of 3.5 inches. The bob is partially hollow at the top to allow the threaded cap to be screwed into it, and partially hollow at the bottom to allow the threaded tip to be screwed in as well.

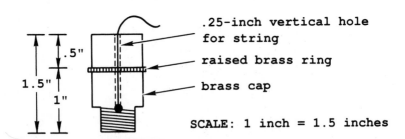

Figure 2-1: Cut View of Threaded Cap and String

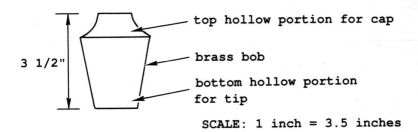

Figure 3-1: Side View of Bob

Threaded Tip

The plumb-bob tip is cone-shaped, reducing to a sharp point at its bottom end. This allows the user to place the end directly over a specific spot on the ground and thereby increase the accuracy of the sight given. The tip is made of hardened steel and is machined shiny and smooth from the threads to the point. A horizontal hole drilled through the tip allows a nail or other piece of metal to be threaded through to help unscrew the tip if it is rusted to the bob. Through use the tip may become worn and dull and should be resharpened or replaced.

Conclusion

Plumb bobs are used in many surveying and construction situations. For example, if you wanted to locate a property corner after having found an iron pipe marking the spot, the plumb bob would provide a sight that the surveyor could use to locate the corner.

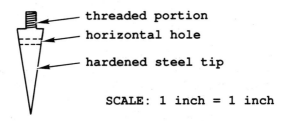

Figure 4-1: Cut View of Tip and Horizontal Hole

The actual operation of the plumb bob is relatively easy. First, stand over the specific point (the corner) so you are facing the surveyor and the corner is between the two of you. Second, hang onto the string with two fingers of either hand, and allow the plumb bob to hang in midair above the corner. Third, spread your legs somewhat so that you are stable, look directly at the point of the plumb bob, checking that it is centered over the corner, and allow the surveyor to take a sight on your string.

●**EXERCISE 1**   Take in hand a common desk supply object, such as a stapler, a hole-punch, or a renewable calendar pad-holder. Look carefully at the object, examining it in the smallest detail. Disassemble it if practical. Try to see in every dark corner. Then list every single part of it and classify the parts into several major divisions.

●**EXERCISE 2**   Write a technical description of a family game that you make up. Pretend you are employed at Parker Brothers and are submitting this technical description in order to have the company accept your idea and produce the game.

●**EXERCISE 3**   Write a technical description of a rather simple object that has a replaceable part, such as a safety razor, and write a conclusion that includes making the replacement as part of the operating cycle.

●43

INSTRUCTIONS

KEY POINTS

- Be sure you address your intended audience.
- Organize your instructions carefully.
- Define your steps clearly.
- Use graphics to help illustrate your material.
- Write all instructions simply and directly.

Technical writers are often called on to give instructions. Many instructions are oral, but many others must be written down—those that are frequently used, those that go to a customer, and those that go to the general public.

Instructions can range from simple ones that are shrink-wrapped with a gadget when it is packaged, to complex, highly technical ones that only a few people understand. Because of the wide use of instructions in the world of technology, it is important to know how to write clear, organized, and precise instructions.

43.1 ANALYZE YOUR AUDIENCE.

Before you begin to write a set of instructions, it is important that you determine for whom your instructions will be written. Ask yourself if your audience is highly technical, semitechnical, or nontechnical. If the audience is nontechnical, determine further if it is a general audience with every type of interest and reading level or if it is one with a high reading level. Ask yourself what your audience will need to know to understand your instructions.

Most instructions are written for the general public and are aimed at anywhere from a third- to sixth-grade reading level. This may seem shocking, but when you consider the short, to-the-point sentences and the simple language of instruction writing, it is believable.

43.2 ORGANIZE AN INTRODUCTION TO THE BODY OF INSTRUCTIONS.

The length of your instructions will determine how complicated your introduction needs to be. For example, the instructions for inserting the picture in the key ring (see Figure 43-1) begin with a two-sentence introduction, for a long introduction is not needed with such a short set of instructions and would simply add unnecessary words to an otherwise clear set of instructions. On the other hand, in the set of instructions for making a college student's pizza (see Figure 43-2), an introduction was needed to explain why someone would want to make such a pizza and to list the needed ingredients and utensils.

You can include a number of items in your introduction—some or all of the following:

- Purpose for the instructions
- Skills needed
- Materials, equipment, and special conditions
- Definitions
- Warnings or cautions
- List of major steps

a. Skills needed

If special skills are needed, it is important that you explain this. A person who is going to follow your instructions needs to know what, if any, training or skill is needed in order to do so.

b. Materials, equipment, and special conditions

Materials and equipment are usually listed if they number more than two or three. Sometimes they are listed separately and sometimes together. Normally, they are listed in the order in which they are used.

c. Definitions

If terms are employed that the user of the instructions might not understand, define them. If no more than three words need defining, you can include the definitions in the body of your instructions unless the terms are used frequently. Then you need to list the definitions on a separate page and refer to it. If you need to define more than three words, list them in a special definition section. Quite often this section is found at the beginning of the instructions, although a few writers prefer to place them at the back. Placing them in the front tells your reader right away that the definitions are available. It is more considerate to present the definitions in the introduction and not as a big surprise at the end of the instructions—after your reader has struggled with your terminology.

d. Warnings and cautions

Although your "WARNINGS" and "CAUTIONS" should appear throughout your instructions, first place them together at the beginning of your instructions to give your reader an idea of what to expect in the instructions. List warnings first and cautions second. See Section 43.3, p. 379, for a complete discussion of the difference between and use of warnings and cautions.

e. List of major steps

If your introduction is long, end it with a brief listing of the major steps you will be presenting in the instructions.

EXAMPLE INTRODUCTION

LAYING COUNTERTOP

by Mark Kahn

You've seen that durable finish on your neighbor's bar, and you appreciate kitchen stove counters, and now you want to put some on that special project of your very own, but you're afraid that it's too difficult for you, and you can't afford a carpenter.

Of course we're talking about countertop. These instructions will explain to you the steps for laying down almost any type of countertop. If you have ever used a few power tools, such as a table saw and an electric sander, and if you have ever worked with wood before, you will discover how easy it is to lay your own countertop.

43.2

Materials and Equipment

First, take a moment to look through the list of materials and equipment that you will need:

Portable circular saw with plywood blade Used to rough-cut pieces out of sheets of countertop.

Electric sander Used to smooth the surface of the project. It may be a belt or orbital type.

Wood fill and putty knife Used to fill in cracks or chips on wood surfaces.

Contact cement (adhesive) Used to bond the countertop to your project surface.

Paint brush Used to apply the contact cement.

Wax paper, loose venetian blinds, or wooden dowels Used to separate the two surfaces of the countertop and your project.

Rolling pin or countertop roller Used to smooth out air bubbles under the countertop.

Router Used to bevel, or smooth, the countertop edges by removing all unwanted overhang.

Cautions

Before proceeding, be sure to read the following cautions.

CAUTION: DO NOT ATTEMPT TO UNDERTAKE THIS PROJECT IF YOU HAVE NOT USED AN ELECTRIC TABLE SAW. YOU COULD DAMAGE THE MATERIALS YOU ARE WORKING WITH.

CAUTION: ALWAYS CHECK FOR CRACKS OR CHIPS BEFORE LAYING COUNTERTOP. THESE CAN CAUSE AIR BUBBLES UNDERNEATH THE NEWLY LAID COUNTERTOP.

CAUTION: MAKE SURE YOUR COUNTERTOP DOES NOT SLIP WHILE YOU REMOVE THE SEPARATORS BECAUSE IT IS IMPOSSIBLE TO MOVE COUNTERTOP WITHOUT BREAKING IT ONCE IT COMES IN CONTACT WITH THE AREA IT IS TO COVER.

The Steps

Now that you are familiar with the equipment, materials, and cautions, look through the steps. They are divided as follows: preparing the surface, applying the adhesive, laying the countertop, and finishing the project.

43.3 WRITE CLEAR INSTRUCTIONS.

The body of the instructions contains the steps for the instructions themselves. They are often developed as follows:

First Set of Steps	**Second Set of Steps**
Definition and Purpose	Definition and Purpose
WARNINGS, CAUTIONS, and NOTES	(and so forth)
Instructions for Step One	

a. Definition and purpose

Often it is necessary to define your steps and to state why they are necessary. The "definition and purpose" are often included in the same sentence. For example, the following set of instructions on surface preparation for laying countertop explains why the preparation is needed.

EXAMPLE

> Surface preparation is the careful removal of all dirt and chips, and the smoothing of all holes, cracks, and depressions with wood fill. This is done on the surface the countertop will adhere to in order to prevent air bubbles from forming underneath the countertop.

b. Warnings, cautions, and notes

Warnings, cautions, and notes are a very important part of any instruction writing. Because of this, you need to understand the purpose and use of each.

Warnings "WARNINGS" are the strongest of the three statements. They warn a user that the product, equipment being used, or the like, is dangerous and could cause serious human injury. The format for warnings varies with different companies, but the content is always very carefully written, partly to protect the user and partly to protect the company against lawsuits. In some companies warnings are used against both serious human injury and serious damage to a piece of equipment. However, most companies use "CAUTIONS" against any type of damage to objects.

A warning consists of three parts: the warning, precautions to be taken, and the remedies in case of accident.

EXAMPLE

<u>WARNING</u>: ENAMEL PAINT IS BOTH TOXIC AND FLAMMABLE.
[the warning]
DO NOT BREATHE VAPORS. USE IN WELL-VENTI-
LATED AREA. WEAR SPLASH GOGGLES, SOLVENT-
RESISTANT GLASSES, AND OTHER PROTECTIVE
GEAR. [the precautions to be taken]
IN CASE OF EYE CONTACT, IMMEDIATELY FLUSH
WITH WARM WATER FOR TEN MINUTES AND SEEK
MEDICAL ATTENTION. IN CASE OF SKIN CONTACT,
WASH WITH SOAP AND WARM WATER. [the remedies
in case of accident]

Note that "WARNING" is underlined and capitalized. Also, the text of the warning is capitalized. This calls attention to it. Further attention is called by placing the title on the left hand side of the page by itself and lining up the rest of the text two spaces beyond the colon after "WARNING."

Cautions Cautions are generally used to caution against damage to equipment and materials. Because it is easy to caution a person against doing something, but not so easy to prevent the person from doing it anyway, cautions normally contain a cause and effect relationship; that is, they explain the reason for the caution.

EXAMPLE

<u>CAUTION</u>: WEAR CLEAN GLOVES WHEN HANDLING CLEANED
PARTS. SKIN OILS CAN CAUSE POOR ADHESION OF
THE ENAMEL.

To tell someone to wear gloves is not very effective, but to explain the reason for wearing the gloves makes a difference, particularly if the person reading the caution has no desire to repaint the object.

Notes Although "NOTES" are generally included with warnings and cautions, they are not as critical to the instructions. Notes give suggestions that can help a person do a better job.

EXAMPLE

<u>NOTE</u>: If necessary, thin with Reducer Number 60 to obtain
proper spraying consistency.

Because it is less important than warnings and cautions, the material in the note is not capitalized, except where needed for standard usage.

Occasionally, a note should be a caution, as in these two examples:

NOTE: Discard any mixture not used within six hours.
NOTE: Mixture has a pot life of approximately six hours.

Neither of these notes is very clear, yet each is trying to caution the reader not to use the product after a certain period of time. Rewritten, the two become one clear caution:

CAUTION: BECAUSE MIXTURE HAS A POT LIFE OF APPROXI-
         MATELY SIX HOURS, DISCARD ANY MIXTURE NOT
         USED WITHIN THIS TIME.

When a series of warnings, cautions, and notes appears together in a text, always place them in descending order of importance: warnings first; cautions second; and notes last. Always place warnings and cautions in instructions *before* a step. If they are placed after a step, the harm may already have been done. Notes can be placed either before or after a step. Use logic to decide where the note belongs.

c. Instructions

Your instructions for each set of steps must be clearly worded. Keep your audience in mind. Write your instructions in the imperative mood (see Section 25). This means that you will be using the "implied you." Such commands as "Open the valve before proceeding" and "Be careful not to break the glass tube," are examples of the imperative mood.

Keep these rules in mind when writing instructions:

1. Include only one step in any one entry in your instructions. You do not want to confuse your audience with too many details crammed together.
2. Do not change the name of an object to provide variety. Whatever you call an object at the start of your instructions should be the same as at the end of them.
3. Use pronouns in referring to objects and ideas only when the reference is entirely clear.

4. Number each step, or use some other form of enumeration (see Section 30.1, p. 184, for examples).
5. When using graphics, be sure to label your tables and figures clearly and to make clear references to them in the instructions themselves.
6. Use figures to clarify material, not to replace text.
7. Always write with your audience in mind. Tell them what they need to know.

The following set of instructions for assembling a saxophone demonstrates how to write simple and direct instructions:

1. Place the neck strap around your neck and tighten or loosen the string by moving the sliding device up or down.
2. Take the saxophone body out of the case and hold it in an upright position.
3. Place the hook on the neck strap into the loop in the middle of the back of the saxophone's body (see Figure 1).
   NOTE: The saxophone should now hang without your having to hold it. You may continue to adjust the neck strap for comfortable playing level.
4. Pull the protective cap out of the top of the body.
5. Apply cork grease to the smooth bottom rim of the neckpiece to ease insertion.
6. Turn the receiver screw (see Figure 1) to loosen the top of the body.
7. Insert the bottom of the neckpiece into the top of the body so that the corked top of the neckpiece is facing you (opposite the side of the bell).
8. Tighten the receiver screw.
9. Loosen the ligature connecting pins.
   CAUTION: REEDS ARE FRAGILE. HANDLE WITH CARE TO PREVENT CHIPPING AND BREAKAGE.
10. Place the flat surface of the reed onto the front of the mouthpiece and hold it in place with your thumb.
    NOTE: The reed should fit perfectly over the top opening on the mouthpiece.
11. Carefully, so as not to slip the reed, place the ligature over the mouthpiece and reed so the bottom of the ligature comes just above the base of the mouthpiece.

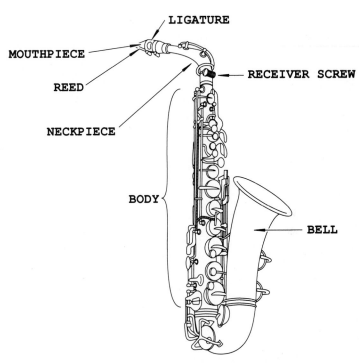

Figure 1. Main parts of an alto saxophone.

Source: Clayton H. Tiede, <u>Practical</u>
<u>Band Instrument Repair Manual</u>,
3rd edition, p. 85.

NOTE: The connecting pins should be on the right side of the ligature. The ligature will not fit correctly with the connecting
pins on the left side.

12. Tighten the ligature connecting pins.

13. Apply cork grease to the corked top of the mouthpiece.

CAUTION: GENTLY TWIST THE MOUTHPIECE ON. DO NOT FORCE
IT, OR DAMAGE MAY RESULT.

14. Twist the bottom end of the mouthpiece onto the corked top of the
neckpiece until secure.

NOTE: The top side of the mouthpiece, the side with the reed,
should be facing down.

43.4 CONCLUDE INSTRUCTIONS WITH NECESSARY INFORMATION.

Your conclusion may be very short if your instructions are short. On the other hand, a long set of instructions needs a longer conclusion. The types of material most often found in conclusions are as follows: Summary of Steps, Interrelationship of Parts, and Troubleshooting.

a. Summary of steps

In a long set of instructions you may want to conclude with a summary of the steps you have covered. For example, the instructions on laying countertop end with a summary.

EXAMPLE

Laying countertop involves four basic steps: preparing the surface, applying the adhesive, laying the countertop, and finishing. In preparing the surface, be sure the surface is smooth and clean. When applying the adhesive, be sure to use the correct adhesive and allow it to dry for no longer than 15 minutes or until "tacky." In laying the countertop, work quickly so the countertop is not inaccurately adhered to the surface. Finally, using a router, bevel the edges.

b. Interrelationship of parts

Sometimes it is necessary to show how the various parts of an item interrelate. For example, in order to operate some equipment, it is important to know how the process works as a whole.

c. Troubleshooting

Often instructions will conclude with a troubleshooting section to inform the reader of what to do if a piece of equipment is not operating properly. This section is frequently placed on a separate page for easy access.

43.5 USE GRAPHICS TO CLARIFY STEPS.

Many instructions would be very difficult to follow if they were not accompanied by clear graphics. Section 39 discusses the various types of graphics that are useful in technical writing. Choose the ones that best fit your manual.

In the examples of instructions that follow, Figures 43-1, 43-2, and 43-3, notice how graphics are incorporated into the instructions.

FIGURE 43-1   SHORT SET OF INSTRUCTIONS EMPLOYING A DRAWING

INSTRUCTIONS FOR INSERTING A PHOTOGRAPH IN THE HALLMARK PLASTIC KEY RING

The Hallmark plastic key ring is a heavy plastic disk that holds a 3-1/2-inch-diameter photograph inside its lucite cover. To insert the photograph of your choice, follow these simple directions:

CAUTION: WHEN REMOVING THE COVER, DO SO GENTLY SO AS NOT TO CRACK THE PLASTIC DISK.

   NOTE: It is easier to insert the photograph when the key ring is not attached to the chain.

1. Remove the lucite cover by pressing very gently on the center of the plastic backing. To do this, rotate the disk carefully while continuing to press on the center of the plastic backing until the cover is loosened and can be removed (see Figure 1).

2. Using the lucite cover as a pattern, carefully center it on your photograph and lightly trace with a pencil a circle the size of the cover around the photograph you wish to insert.

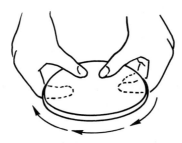

Figure 1.   Pressing and Rotating the Disk to Remove the Cover.

NOTE: Be sure you trace a large enough circle so that your photograph will fit tightly in the disk.

3. Carefully cut out the photograph with a sharp scissors or a craft
   knife.
4. Place the photograph face up in the plastic disk.
5. Gently reinsert the plastic cover over the photograph.

FIGURE 43-2 LONGER INSTRUCTIONS WITH GRAPHICS

BREAD DOUGH PIZZA FOR THE COLLEGE STUDENT

Pizza is enjoyed by people of all ages. Unfortunately, a simple pizza
that feeds four can cost fifteen dollars or more. College students can't
afford these prices, but can fix easy-to-make, great-tasting pizzas,
which easily feed four people for around five dollars.

To make such a pizza, gather the following ingredients and
utensils:

INGREDIENTS	UTENSILS
1 loaf of frozen bread dough	1 medium-size fry pan
1/2 lb hamburger	1 spoon
2 tablespoons oil	1 strainer
1 small jar pizza sauce	11" by 20" by 1" cookie pan with sides
1 teaspoon oregano	
1 pound grated mozzarella cheese	

Little or no cooking experience is needed to make this delicious
pizza, which is made in five easy steps: thawing the dough, browning
the meat, preparing the crust, making the topping, and baking the
pizza.

Thawing the Dough

Remove bread dough from freezer and thaw it for 4 to 5 hours at
room temperature or for 24 hours in a refrigerator.

Browning the Meat

1. Place 1 tablespoon oil in a medium-size fry pan and bring it to a medium heat.
2. When oil is hot, place hamburger in it and brown it by breaking it apart with a spoon into small pieces.
3. Season with salt and pepper to taste.
4. When hamburger is completely browned, place it in the strainer to drain excess grease.
5. Allow hamburger to cool in strainer while preparing the crust.

Preparing the Crust

1. Preheat oven to 400 degrees.
2. Grease the bottom and sides of cookie pan with 1 tablespoon of cooking oil.
3. Squeeze dough between hands until air bubbles are removed.
4. Place dough in center of cookie pan and spread it out by pressing dough down with both hands together and then spreading hands apart (see Figure 1). Allow dough to fall over the edge of the pan by 1 inch.

   NOTE: The dough will tend to pull back together; therefore, keep working the dough until it remains spread out.

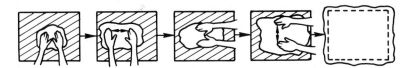

Figure 1.   Placing the Dough on the Cookie Pan.

5. Roll the excess dough edge over to form a ridge as shown in Figure 2.
6. Pour pizza sauce on the dough and spread it over the entire area of dough.
7. Sprinkle 1 teaspoon of oregano seasoning on the sauce.
8. Make a layer of cheese by spreading approximately two-thirds of the grated cheese over the sauce.
9. Take the browned hamburger and spread it over the layer of cheese.
10. Spread the remaining cheese over the layer of hamburger.

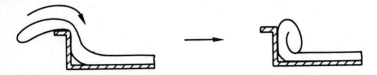

Figure 2. Forming Ridge on Edge of Cookie Pan.

Baking the Pizza

1. Place pizza in preheated 400-degree oven for 15 to 20 minutes or until cheese is brown on top.
2. Remove from oven and allow pizza to stand for five minutes.
3. Cut into slices and serve.

These simple steps will produce a mouth-watering and inexpensive pizza to enjoy with friends or roommates.

The instructions on the following pages are one section of a large and detailed training manual for a building supervisor of a university student center.

FIGURE 43-3 A SECTION OF A TRAINING MANUAL

CLOSING THE GAME ROOM

The Game Room closes at

11:00 pm	Sun, M, T, W, Th
12:00 am	F, Sat

You should go down to the Game Room about ten minutes before its scheduled closing time. Take with you the following things:

- Pinball bags
- Pinball slips
- Pinball keys

PINBALL MACHINES

The first thing to do when you get down to the Game Room is to close the pinball machines.

- Fill out the pink pinball slips.

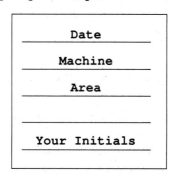

- Collect the money from each machine and put it and the corresponding slip into a pinball bag.
- Lock the pinball bags with the locks that are inside them.

CASH REGISTER

Next you will Z-out the cash register. This will zero all of the memory in the register so it will be ready for the next day.

- Insert your register key into the register and turn it to the supervisor position.

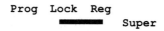

- Press the $\boxed{\begin{array}{c}\textbf{Jnl}\\\textbf{Feed}\end{array}}$ key three times.

- Press $\boxed{2}$ $\boxed{\begin{array}{c}\textbf{Err}\\\textbf{Corr}\end{array}}$ $\boxed{\textbf{Total}}$.

- When the register stops, press $\boxed{\begin{array}{c}\textbf{Jnl}\\\textbf{Feed}\end{array}}$ and $\boxed{\begin{array}{c}\textbf{Rec}\\\textbf{Feed}\end{array}}$ each three times.

- Tear off the receipt tape and fold it up.

- Take out the journal tape and fold it up.
- On the bottom of both tapes write:

```
Date
Day
Time
Area
Your Initials
```

- Put both of these tapes into the Game Room fund bag.
- Enter the next day's date into the register by pressing the corresponding keys.

    Feb. 2, 1987   press [0] [2] [0] [6] [8] [7]

- Press the $\dfrac{\text{H.O.}}{\text{Date}}$ key.
- Turn your key to the lock position and remove the key.
- Turn the register off.

<u>NOTE:</u> Leave the cash drawer open, as this shows potential thieves that no money is left in the till.

<u>THE VALIDINE</u>

Next, turn the validine machine off.

- Press [Control] [2]
       [Control] [3]
       [Control] [±] [1]
- The machine will now read "TAPE OUT."
- Remove the tape.
- Unplug the phone jack first and then the wall plug.
- Cover the machine.
- Place the tape in the tape box located in the top drawer of the file cabinet in the back room.

LOCK DOORS AND GATES

The following doors should now be locked:

- Swinging door to the Game Room
- Security gate to Game Room
- Door in the back room
- Far door to the bowling alley

LIGHTS

In the back room, by the pool cues, are the light switches. Turn them all off except the blue ones that read "Exp."

PINBALL MACHINES OFF

In the back room, above the desk, is the breaker panel. Switch off the three breakers just below the one that reads "Leave Off."

		Leave
On	▭▬	Off
On	▬▭	Off
On	▬▭	Off
On	▬▭	Off

CHECKLIST

Just before you and your Game Room attendant walk out of the door, run down this list in your head.

- ☐ Are all necessary lights turned off?
- ☐ Are all doors and security gates locked?
- ☐ Is the cash register turned off?
- ☐ Is the cash drawer open?
- ☐ Is the validine unplugged?
- ☐ Is the validine tape in the file cabinet?
- ☐ Do you have your keys?
- ☐ Do you have all of the pinball bags?
- ☐ Do you have the pinball keys?
- ☐ Does your Game Room attendant have his/her keys?
- ☐ Does your Game Room attendant have his/her fund bag?

NOTE: If your answer to any of these questions is no, go back and fix it.

If the answers to all of the questions on the checklist are yes, then walk out of the remaining bowling-alley door and lock it behind you.

DROPPING THE FUNDS

You and your Game Room attendant walk up to Accounting.

- Unlock the drop box, which is located just inside the Accounting door, and drop the Game Room fund bag and the Game Room keys into it.
- Close the drawer and make sure that you hear the contents fall into the safe.
- Initial the drop sheet, which is on the clipboard on top of the counter, next to where it says Game Room. Also have your attendant initial the sheet.
- Read off the numbers on the pinball bags and have your attendant check them off on the sheet. Then drop them into the drop box.
- Lock the drop box.

●**EXERCISE 1**   Collect as many examples of instructions as you can and bring them to class. Compare yours with those of other students. How clear are the instructions? Do any of them need to be rewritten?

●**EXERCISE 2**   Select a short set of instructions from those that you brought to class. Rewrite them to conform to the suggestions for clear instructions outlined in this section.

●**EXERCISE 3**   Write your own set of instructions. Some suggested topics follow:

How to decorate a wedding cake
How to make a meatloaf
How to change a flat tire
How to build a campfire

How to install a wood-burning stove
How to put out a kitchen fire
How to change the oil in your car
How to use a simple software package
How to play a game, such as checkers

●44

PROCESS ANALYSIS

KEY POINTS

- Make a clear distinction between process analysis and instructions.
- Choose introductory material according to the needs of your reader.
- Separate the steps of the process into clear major divisions.
- For the most part, write brief conclusions.

A process analysis explains how something is done or produced, how something works, or how something has occurred. In the first instance—how something is done or produced—process analysis seems to overlap with instruction. The difference is an important one. Instruction teaches a person actually to perform a task; process analysis teaches a person to understand a process, not to perform it.

Process analysis plays an important role in technical communication. One of its main uses is for manual writing. Technicians and operators often read process explanations in order to understand better the actions they must actually perform. Supervisors may never perform certain tasks that they oversee, but they read process explanations for the understanding necessary to do their job effectively. Sales materials often explain the processes of the products they promote: marketing personnel must understand their product in order to sell it, and customers must read the explanations in order to make wise purchases.

Process analyses can discuss processes in which human activity plays a large part, such as the way glass is blown. They can focus on large-scale processes, such as the operation of a cyclotron, allowing individual workers to perform their tasks with comprehension of the overall process. Process analysis can also explain a subject in which humans play little or no part, such as the formation of volcanoes, geological layers, or solar energy. These processes cannot be controlled but must be understood in order for people to do such things as drill in the right place, order suitable equipment, design effective products, or build in a safe location.

44.1 ANALYZE CAREFULLY WHO IS GOING TO READ THE PROCESS EXPLANATIONS.

Audiences for process analyses include the general public; students, trainees, and first-time users; customers and potential customers; and supervisors and managers. As always, you must know your readers' grasp of the subject and need for the explanations, or for any part of them, before you can make appropriate choices about the inclusion of material and the style. You have to decide how many and what kind of graphics to include, and whether to discuss special equipment and the conditions necessary to the process. You also must decide the amount of definition and theory, as well as the technical level of your vocabulary and the sentence/paragraph lengths. It is usually safe to assume that at least some of the readers are nontechnical and inexperienced in that particular process. It is possible, however, that the process analysis has an audience that already has been working in some related aspect of production or sales, so that some material may be excluded and the style may be somewhat more technical.

44.2 UNDERSTAND YOUR AUDIENCE'S NEED FOR THE PROCESS ANALYSIS.

The process analysis may be accompanied by a set of specific instructions, making it an integral part of a worker's ability to perform a task. Ask yourself questions, such as, is the function to explain the sophistication of a potential purchase in order to justify the cost? Is the process analysis going to help a salesperson explain a process to a potential customer? Is an experienced technician in need of a better understanding of the overall process of which he or she is a part in order to understand changes in the process or new equipment? Does a manager need to understand the process in order to tell others what their work means to the company and to manage their work effectively? Does management need to understand the working of a hydroelectric power dam, for example, in order to know whether to build a radio station in its vicinity? Does a homeowner require an understanding of solar cells before investing in a solar-energy home? Your decision about an audience's need for an explanation determines what parts to emphasize and what types of introductory materials to include.

44.3 GAIN COMPLETE UNDERSTANDING OF THE PROCESS BEFORE YOU BEGIN TO WRITE.

If you can personally observe or even perform the process, you should do so. Whether or not this is possible, you will doubtless use the library to gather historical and background information, as well as more complete knowledge of the principles and theories involved, so that your information and interpretations are accurate. Library research is a skill in itself; the investigation of secondary sources plays an integral part in the composition of many process analyses.

The questions you should answer follow:

1. What is the process?
2. What is its function?
3. Where and when does it take place?
4. How does it work?
5. Why does it work?
6. Who or what performs it?
7. What are the principle steps involved?

44.4 USE THE INTRODUCTION TO GIVE THE READER THE NECESSARY OVERVIEW OF THE PROCESS.

Your introduction section can be very lengthy or it can be one paragraph (or even one sentence) long, depending on the audience and the purpose of the explanation. Most process analyses give at least a definition of the process and state the paper's purpose—and perhaps its audience. The introduction also may include several other types of information in order to orient the reader, who must be ready to receive and understand the process analysis that follows it. Introductions include some or all of the following information:

1. Definition of the process
2. Purpose and audience of the process analysis
3. Historical background
4. Overriding theories and principles of operation
5. Future considerations that involve the process
6. Special considerations that involve the process
7. Equipment and supplies needed for the process
8. Current operators of the process
9. Major steps of the process

EXAMPLE

CONTINUOUS-PROCESS DIP PAINTING

Definition and Purpose

Continuous-process automated dip painting is an industrial finishing process that has a minimum of manual employee involvement and offers the maximum protection for the parts painted and maximum uniformity of the finish. Capital outlay is high, but per-unit cost is low for high-volume operations. Expanding manufacturing facilities should consider this process for uniformity and cost effectiveness.

Background and Theory

This process is a child of the industrial revolution. It would never have arisen in a world of small job shops. It is extensively employed by the automobile manufacturing industry, including their sub-contract suppliers.

The principle that doubtless was the germ idea is that when an object is immersed, it gets wet all over. Thus the perennial problem of unpainted areas is avoided.

Special Considerations

The process as it will be described here is tailored for the painting of steel. Variations are employed for the painting of other materials. For example, the vapor-phase degreasing cannot be used on assemblies containing plastics.

Equipment and Supplies

The process requires a very long, motor-driven overhead conveyor, larger-than-usual gas-service line, three large bevel-ended immersion tanks, and several hundred infrared lamps.

Also required are several hundred gallons of trichloral ethylene solvent, and about a ton of zinc chromate. The latter is consumed in the process, while the former, in theory, is never used up, but, in fact, at least a gallon is lost for each hour of operation.

List of Principle Steps

Continuous-process dip painting involves four major steps. First, the parts must be manually suspended from the conveyor; second, they pass into the vapor-phase degreasing tank; third, they pass into a liquid solution of zinc chromate; fourth, they are lowered by the conveyor into a tank of paint; fifth, from there they pass between long rows of infrared lamps; and sixth, they are removed from the conveyor.

44.5 DIVIDE THE BODY OF THE PROCESS ANALYSIS INTO MAJOR CHRONOLOGICAL STEPS TO FORM AN EASY-TO-COMPREHEND NARRATIVE.

a. Classify the detailed steps of the process into major steps.

Before you write the body of the paper, you should list all of the steps in the process. Then classify these individuals steps into major ones. Try not to make too many or too few major divisions.

b. State your major steps in parallel phrases or sentences.

Look at the previous analysis of continuous-process dip painting and notice that the major steps are all stated with like or parallel grammatical forms of expression. To achieve parallelism, each step was given the same parts of speech. Other examples of parallel wording follow.

EXAMPLES

emptying the gun	the mixture of the soil
preparing the gun	the transfer of the plant
cleaning the gun	the nature of the transfer
using the gun	the health of the plant

c. Put just enough detail in each step to satisfy your reader's needs.

If your reader does not need to know a vast array of detail, and if, moreover, such detail would clutter the narrative, then leave much of it out. If your reader knows well the basic outline of the process, you may need to provide clarifying details. Usually this audience is more technical, but not necessarily. A nontechnical reader may need much detail in order to understand a process at the required level of, say, a supervisor.

d. Divide complicated major steps into substeps.

If the subject is complex, the details within major steps may need organization, also. You can announce substeps in the introduction to the major steps and then discuss them in the chronological order of the whole process. For example, in a process analysis on obtaining an engineering degree, the first major step is a discussion of the university's requirements, which break down into several categories of study, and the second major step is the study of professional requirements, which break down into emphases within the major.

Note: No matter how much division you do, always keep the material in chronological order.

EXAMPLE

These steps are from the Continuous-Process Dip Painting analysis.

Loading the Conveyor
Unskilled factory workers attach the parts to the conveyor by means of hangers. Their only concern is that all parts have the same orientation, so that the small bubble of excess paint will always be at the same place on the finished product.

Degreasing the Parts
Vapor-phase degreasing takes place in a tank about twenty feet long and only two feet wide and six feet high. Large gas burners below the tank continuously boil a layer of trichloral ethylene. The vapors

rise to a level near the top of the tank where a water jacket surrounds the tank. The water cools the vapors, causing the formation of a cloud continuously raining the trichloral ethylene back into the tank. As the room-temperature parts are lowered through this cloud into the vapor region, the hot vapors condense on the parts, heating them up again, and then the vapors fall back to the bottom of the tank as liquid. The liquid carries with it all grease from the parts.

Preparing the Surface

The conveyor now lowers the parts into a tank of zinc chromate kept hot by the parts themselves. As the parts immerse, a layer of zinc chromate adheres to the steel. Paint will stick to the chromate better than it would to uncoated steel.

Painting the Surface

The parts now pass down into another 20- by 2- by 6-foot tank containing paint. The parts come back up, rising above the paint quickly, but remaining over the tank long enough to let the paint drip back into it.

Drying the Parts

The conveyor now passes between long rows of infrared lamps that speed the drying of the paint.

Removing the Parts

After the parts are painted and dried, they have only to be removed from the conveyor by workers wearing gloves to protect their hands from the heat.

44.6 KEEP THE CONCLUSION BRIEF.

Usually the concluding statement in a process analysis is a short paragraph or two summarizing the major steps. If the paper is very long, the conclusion may well be longer and you would probably discuss any implications of the process for people in general, the company, and so on. You also might stress some part of the process as particularly important.

EXAMPLE

Conclusion
    Continuous-process dip painting has several advantages in the
areas of cost and uniformity over the chief competing process, spray
painting. By comparison, spray painting is labor intensive, wasteful of
material, and produces a product lacking uniform quality.

44.7 USE GRAPHICS AS THEY ARE NEEDED TO CLARIFY AND ENHANCE THE NARRATIVE.

    Graphics in a process analysis serve a special function in demon-
strating the dynamics of motion in diagrams and other visuals of a
process (see Section 39).

44.8 WRITE IN ACTIVE VOICE AND THIRD PERSON FOR PROCESS ANALYSES.

    Active voice, where the subject of the sentence does the action, is
always the preferred voice; use the passive voice, where the subject
receives the action, only on occasion (see Section 26). When possible, describe the process without reference to *I* or *me* or *you* or *us*,
or even the implied *you*, such as is common in instructions (for example, "Open the small door and latch it in place.")
    The following example is of a complete process analysis paper.
Compare this with the same subject handled as a set of instructions
(see Section 43).

EXAMPLE

THE LAYING OF COUNTERTOP

Introduction
    Countertop is a type of work surface that is scratch-resistant,
highly durable, and made of a mica-like material. As a work surface,
countertop (the brand name *Formica* has become an accepted generic
name for this kind of surfacing) is very desirable because it can take so

much abuse and show no wear. Any homeowner who undertakes to resurface a counter or to purchase a work surface should consider buying countertop.

This hard, smooth surface material is constructed by impregnating special paper with synthetic resins, such as melamine, then subjecting it to heat and pressure; about seven sheets bond together to form a hard and durable material about one-sixteenth inch thick. The top sheet is always patterned and usually colored. The back side adheres to plywood or other suitable backing.

When a person applies countertop, several pieces of equipment are necessary: a portable circular saw with a blade for cutting plywood, an electric sander, wood fill and a putty knife, contact cement, a paint brush, wax paper and wooden dowels, a rolling pin, and a router.

The major steps are preparing the surface, applying the adhesive, laying the countertop, and finishing.

The Process

Surface preparation. Countertop usually comes in large sheets that the homeowner must cut to the size desired. This is done with a standard portable circular saw with a plywood blade. Although the edges won't be smooth at first, they can be routed later.

The underside of the countertop receives the guide marking for the cutting. The cut occurs about a quarter inch away from the line, which leaves room for error.

Only a clean, unchipped surface allows the cement to adhere. Cracks and chips cause air bubbles to form underneath the cement. The heads of nails and screws are always below the surface, and any indentation where they exist is filled in with wood fill. A completely clean, dry surface receives final touches with a sander for a totally smooth surface.

Applying the adhesive. A quality contact cement makes the best bond. Adhesive smoothly covers the entire surface, as it does the bottom of the countertop. When the adhesive has dried to the tacky stage, about fifteen minutes, the countertop is ready for placement. If the adhesive becomes too dry, it will not bond.

Laying the countertop. The best method for preventing accidents involves covering the tacky plywood surface with some kind of protective material, such as wax paper on wooden dowels. This separator is slowly withdrawn as the countertop carefully comes down on the surface being refinished. A rolling pin rolled over the new surface forces out air bubbles trapped between the layers. The process demands precision, for the bonding takes place immediately.

Finishing. After the contact cement solidifies, a router removes the protruding edges of countertop and bevels the edges.

Conclusion

Laying a countertop requires precision cutting and placement of the layers. The process has some forgiveness in it, however, since the router cuts away the excess.

●**EXERCISE 1**   Write a process analysis for any of the following:

Choosing an astronaut
Grafting a plant
Operating a kiln
Making a hologram
Loss of the stratosphere
How a volcano erupts
How a computer works

●**EXERCISE 2**   Find out about your school's enrollment process and write a process analysis explaining to students what happens to their enrollment cards after they turn them in.

● 45

MANUALS

KEY POINTS

- Before you plan your manual be sure you understand your audience.
- Organize and outline your manual carefully before beginning to write.
- Be sure your manual is coherent, clear, and developed.
- Use graphics throughout your manual to clarify the text.
- Edit fully and carefully.

Countless manuals are produced each year. Some are in-house (within the company) documents while others go to customers. Manuals are as varied as any type of technical document. Some are short, perhaps only a few pages in length; others are huge books that are thousands of pages long. Some are very technical; others are written for a general audience. Some of the categories most often used include

Safety	Job Training	Operation
Computer	Software	Repair and Maintenance (or
Do-It-Yourself	Appliance	Assembly, Disassembly, and Repair)

People often complain about the poor quality of the manual that accompanies a product they purchased. What most people do not realize is that an effective, clear manual is difficult to write, mainly because it addresses a variety of readers, none of whom the manual writer can see. For example, a manual for a software program may be read by a computer expert, and it also may be read by a computer novice. In addition to having a variety of readers, manuals often must cover a great deal of material in a limited space. As a result, the manual writer must be able to condense material and still present it as clearly and as thoroughly as possible.

This section discusses some of the features of well-written manuals and provides examples for guidance in writing more effective manuals. You will find much of the material that goes into a manual described in Sections 41 through 44, for manual writing is based on definition, description, instructions, and process analysis. To understand these four methods of development is to understand the basic parts of a manual.

45.1 DEFINE THE AUDIENCE CAREFULLY BEFORE BEGINNING TO WRITE.

For all technical documents, it is important to know your audience. But unlike a letter, a formal report, or almost any type of technical document, a manual is for an audience you will have little or no contact with. Thus you will know less than usual about who will be reading your manual. You may write an appliance manual that will accompany every appliance of its kind that is sold. This means that the manual might be read by a nontechnical person with a third-grade education or by a professional person with a doctorate in physics. As a result, you have the most difficult audience of all: anyone and everyone.

Before beginning even to plan your manual, ask yourself what common characteristics your audience will have and what *every* member of your audience will need to know. This is a difficult task that most writers do not like to spend enough time on, yet it is crucial to producing a good manual.

If possible, specify in the introduction who the intended audience is. For example, the following introduction appeared in a maintenance manual for a French horn with rotary valves:

> This manual is intended for the horn player who has at least six months' experience with the instrument. It assumes some

knowledge of the instrument and does not define any of the terms used.

In contrast, a manual for the assembly, disassembly, and cleaning of an E♭ alto saxophone addresses a different audience.

WELCOME TO MUSIC! You have just purchased an E♭ alto saxophone. In the following weeks, you will be taking lessons to learn how to read music, how to get a sound out of the instrument, and how to execute more advanced music techniques. But as a beginning musician, you not only need to learn how to play the saxophone, you must also learn the basics of assembly, disassembly, and cleaning. These will ensure the correct and most efficient working order of the instrument.

Notice how much more detailed the introduction is for the beginning saxophone player than for the French horn player with some exposure to the instrument. Clearly, both writers analyzed their audiences before they wrote their manuals.

45.2 GATHER ALL NECESSARY MATERIAL.

You probably will be very knowledgeable about the subjects of the manuals that you are going to write. If you were not, you would not be the person writing the manuals. You understand the material, but you must stop and ask if your audience does. You may need to find background information that you do not have at your fingertips. Sometimes you may need to use your school or company library; sometimes you will gather material from other sources, such as technicians who have worked on a project, previous users of a similar product, or former company manuals for similar uses or older models.

Once you have gathered the information you need, you must develop a rough outline. One method for organizing your material for your rough draft is to place your information on note cards. (See Section 40 for a more complete discussion.) The major headings on your note cards can serve as headings for your table of contents, if you have one. If not, they can still help you organize your manual with great ease, for you can place your cards in the order you will need them. As you work from them, you will find that the manual becomes easy to write.

45.2

Whatever method you employ for recording your information, you need to outline it in some way before you begin the manual. The major headings of your outline, like those of the note cards, provide the major headings for your table of contents.

Figure 45-1, the table of contents from the maintenance manual for the French horn, clearly shows the major areas the writer had to gather information for and outline before beginning to write.

FIGURE 45-1 EXAMPLE OF A TABLE OF CONTENTS FOR A MANUAL

Care and Maintenance of a French Horn with Rotary Valves

TABLE OF CONTENTS

45.3 ORGANIZE THE MANUAL BEFORE BEGINNING TO WRITE.

If you already have information for your manual clearly arranged on individual note cards, organizing your manual will be much easier. Remember to consider your readers' needs. Look through Figure 45-2, a student's "Budget User's Manual," and notice how the "Table of Contents" is organized. What is most important is that the manual is arranged in a logical order for the reader. By planning this organization *before* the manual is actually written, you will find that writing the manual will be easier and you will spend less time revising. After you place the operations in logical order, arrange them in numerical sets. Figure 45-2 demonstrates well-organized sets of instructions.

45.4 WRITE THE MANUAL CAREFULLY.

You must write your manual very carefully. Use the third person and indicative mood (see Section 25) for all informative sections. Use the implied *you* and second person (see Section 43) for all directions and instructions.

Do not try to include too much information on a page. Your reader can comprehend only a limited amount of material at one time; therefore, it is best to limit yourself, if possible, to a *maximum* of three numbered sets of instructions per page. Leave plenty of white space around instructions and graphs. Actually, white space makes both your print and your graphics appear larger than they really are.

Use enumeration (see Section 30) to set off instructions and important points so that your reader can comprehend them more easily.

Divide your manual clearly. Include a front section, an introduction, a body, and a conclusion. Always remember that your main objective in your manual is to convey your information as clearly and accurately as possible.

a. Front section

The front section includes such items as glossaries; lists of abbreviations and symbols; warnings, cautions, and notes; and a table of contents, depending on the needs of your manual.

Include glossaries in front when you have a number of terms that need defining and you want your reader to be able to locate the definitions easily. The glossary in Section 41.3, p. 363, is included in the front of a manual that introduces new students to the computer center at a university. Glossaries are sometimes included in the conclusion of a manual, but they are often buried there. By placing a glossary at the beginning of a manual, you let your reader know right away which terms are defined and where they are located for easy reference.

Include lists of abbreviations and symbols only in manuals where you are using many of them and the manual would be cumbersome to read if they were written out. If all of your abbreviations will be clear to every reader, such a list is not needed. As with definitions, abbreviations and symbols are best placed at the beginning of a manual for the sake of your reader.

One further consideration is whether your manual will be translated into other languages. If it will be, you must be sure that both your abbreviations and symbols and your terminology are universally known—as easily recognized in another country as they are in yours. Your translator must be able to understand what you write. If you are unsure, you definitely will want lists of nomenclature, symbols, and abbreviations—all clearly defined.

If you are using a number of warnings, cautions, and notes, you should first place them in a section in the front part of your manual. In this way your reader is forewarned. (Section 43 has a lengthy discussion of their proper use.)

Use a table of contents if your manual is long enough to warrant one. They usually are put on a separate page and vary in length.

Sometimes it is necessary to include a brief informal introduction to give your reader information before beginning the manual. The two introductions in Section 45.1, p. 405, are examples of introductions that identify the audiences for the manuals. A manual on how to regrip a golf club has the following introduction for a different purpose.

This manual is written so that you can easily regrip golf clubs. This manual is limited to golf clubs with metal shafts because extra knowledge is needed when regripping fiberglass and wooden shafts.

In two sentences the writer states the purpose of the manual and explains the manual's limitations and the reason for them.

Finally, number the front matter of a manual with lowercase Roman numerals (i, ii, iii, iv, v, and so forth).

b. Introduction

The introductory material in a manual can include job descriptions, materials and equipment lists, descriptions of mechanisms, or a more detailed introductory statement that clearly identifies an audience or the manual's purpose.

Job descriptions are placed at the beginning of manuals so trainees can read them first and then know where to find them.

Equipment and materials lists are common at the beginning of manuals. Often these lists include graphics; some include information about where to buy materials. In the following example, the equipment is illustrated for the user.

EXAMPLE

NEEDED EQUIPMENT

The following diagrams show the equipment needed to regrip a golf club. Most of the equipment can be found at a hardware store, with the exception of the level-action shaft holder, the pressure grip remover, and the new grip. These can be purchased or ordered at a golf shop or a sport shop. The tray will be used to catch gasoline, so an old tray will do.

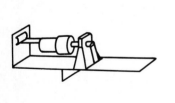  

1. Level-action   2. Pressure grip remover   3. New grip
   shaft holder

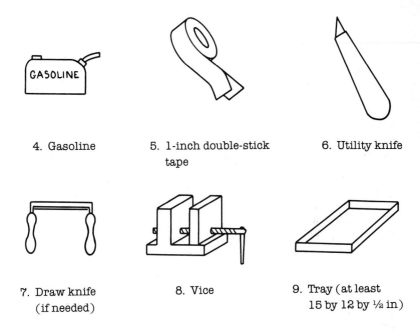

4. Gasoline

5. 1-inch double-stick tape

6. Utility knife

7. Draw knife (if needed)

8. Vice

9. Tray (at least 15 by 12 by ½ in)

If you wish to include a detailed description of a mechanism in your manual, you normally do so in the introduction.

c. Body

Divide the manual's body into as many sections as you need. Under each section give explicit instructions for the steps. Refer to Section 43 for complete details on writing clear instructions.

As a manual writer, you will have many choices in constructing the body of your manual. Your main concern will be the clarity of your material. Consider the following:

Order   Present your material in logical sequence. For example, the writer of the software training manual included at the end has a built-in logical order to follow. By contrast, writers of operation manuals have to decide which order would best serve the readers.

Details   Include all necessary details. Simplify your instructions and then fill them in with all the details that you know your audience needs to understand the material or to follow the instructions you are giving them. Figure 45-2 makes excellent use of details.

Clarity   Without clarity, your instructions will be impossible to understand. Be sure each sentence is stated as clearly as possible. Do not assume that what is clear to you will be clear to your reader.

d. Conclusion

Although conclusions are not essential in a manual, if your instructions are long, be sure to write a conclusion that clearly summarizes your material. Other considerations for your conclusion are troubleshooting, warranties, addresses to write to for further information, and schematics and other materials for audiences more technical than those addressed in the body of your manual.

Occasionally, you might even wish to include appendixes to give your readers further information, such as schematics and other technical material. (See Section 36 for a discussion of appendixes.)

45.5   EMPLOY GRAPHICS WHEREVER THEY ARE NEEDED.

Use graphics generously. Notice how the writer of the manual in Figure 45-2, pp. 414–34, made use of graphics to clarify his material. This is an important aid to readers. (See Section 39 for a discussion of graphics.)

45.6   CAREFULLY EDIT THE MANUAL.

In large companies, many people with a variety of specialties will edit a manual before it goes to a customer. Some companies even have large manual-writing divisions that employ dozens or even hundreds of people. In these places, manual writing has almost been perfected to assembly-line technique. Some people work on nothing but warnings, cautions, and notes; others edit; still others work on one section of a manual, for example, a parts list for a repair manual. Of course, graphic artists provide all the graphics, based on the information supplied to them by the manual writers.

In small companies, employing a large group of specialists is too costly. Thus, it becomes the burden of a writer and a graphic artist to produce the best manual possible. Whichever the case, for the writer, editing is crucial and should be considered as important as the initial audience analysis itself.

45.6

●**EXERCISE 1**   Collect a number of manuals. Try to find different types: computer, repair, appliance, and the like. Bring them to class. Get into small groups and compare your manuals with those of the other members of your group. Decide which manuals are clear and readable and then try to determine the components of a good manual.

●**EXERCISE 2**   Write a manual using one of the topics listed below, or a topic approved by your instructor. Begin by analyzing your audience, proceed by gathering your material and placing it on note cards. Once you have organized your note cards, outline your manual. Then write it. Have members of your class edit your manual carefully before you hand it in to your instructor.

Suggested Topics

A job training manual for a job you have held

A safety manual for your dormitory

A repair and maintenance or care and maintenance manual for a piece of equipment that you own, such as a piece of sports equipment, a musical instrument, or a bicycle

A simple software manual for a novice

A software manual for a computer expert

A computer manual that explains how to perform a simple task such as changing font on a printer

A do-it-yourself manual for a complex task, such as reshingling a roof, building a deck, or making a simple patchwork quilt

FIGURE 45-2 **EXAMPLE OF A STUDENT'S MANUAL**

BUDGET

user's manual

Jane Hassemer
University of Wisconsin—Platteville
1989

45.6

PREFACE

This manual explains how you can use the BUDGET program that is set up on LOTUS 1-2-3 to organize a budget. It assumes you have an IBM-PC, a DOS disk, and the LOTUS 1-2-3 disk, as well as the prepared ATHLETIC disk, which contains the BUDGET program. It also assumes that you are somewhat familiar with a personal computer, in that you can load the disks and align the printer, as well as do other basic operations of the IBM-PC.

The introduction describes the BUDGET program. You will learn the two basic functions of the program—storing records and computing a running balance. The glossary defines frequently used words in the manual that you may not be familiar with. Chapter 1 deals with loading the disks into the computer. Chapter 2 describes how to add records to a file. Chapter 3 describes how to save files for future reference. Chapter 4 deals with changing records that already have been entered. Chapter 5 describes how to print records, and Chapter 6 tells how to quit the program.

TABLE OF CONTENTS

INTRODUCTION

What is the BUDGET program?

The BUDGET program is a program that has been set up in conjunction with LOTUS 1-2-3. A budget structure has been stored on the disk. The structure includes these categories: sport, vendor, encumbered, expenditure, running total of expenditures, and a running balance. Once the first four of these have been entered, the program automatically computes the latter two.

What do I need?

To use the BUDGET program, you will need an IBM-PC as shown in Figure 1. In addition to this, you will need a DOS disk, a LOTUS 1-2-3 disk, the ATHLETIC disk, which contains the BUDGET program, and a printer.

How do I begin?

The first step in using the BUDGET program begins with Chapter 1, "Start Up."

Where can I find additional information?

To find additional functions of LOTUS 1-2-3, which is the basis of the BUDGET program, you can refer to any LOTUS user's manual.

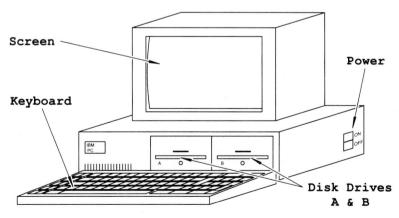

Figure 1. Components of the IBM-PC Personal Computer

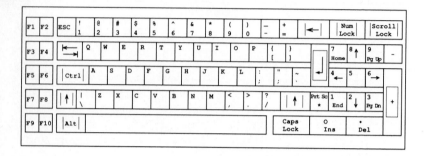

Figure 2. Enlarged View of the IBM-PC Keyboard

GLOSSARY

CURSOR The highlight that shows where you are typing.

DIRECTORY A subdivision of a disk. A directory is a special file that organizes the files stored on a disk.

DISK A permanent storage medium for your material.

DISK DRIVE The device that holds the disks you are using.

FILE A named collection of data stored on a disk.

MENU A series of choices that appear at the top of the screen after you type in the READY mode.

MODE A state in which you can perform a particular process. For example, you can add or change records when in the ready mode.

PROMPT Any message displayed by the disk being used.

RANGE A cell or rectangular group of adjoining cells in the worksheet.

RECORD A set of information, in one row, that contains an entry for each of the column headings on the worksheet.

ROW A horizontal block of cells in a worksheet.

CHAPTER 1

*** START UP ***

<u>WARNING</u>: NEVER REMOVE OR INSERT A DISK WHILE THE RED
LIGHT IS SHOWING. THIS MAY DAMAGE THE DISK OR
RESULT IN LOSS OF INFORMATION.

1. Turn the power off if it is on.

2. Insert the DOS disk in Drive A.

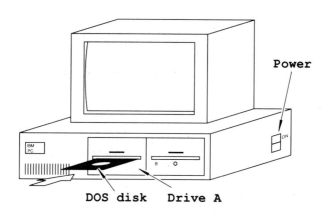

DOS disk   Drive A

Figure 1.1

3. Turn the power on.

   <u>NOTE</u>: Always wait for the red light to disappear before pressing
   any keys or removing or inserting disks.

4. Press ⌐⌐⌐ twice. The A prompt, A>, will appear on the screen, as
   shown in Figure 1.3.

5. Remove the DOS disk from Drive A.

6. Insert the LOTUS 1-2-3 disk in Drive A.

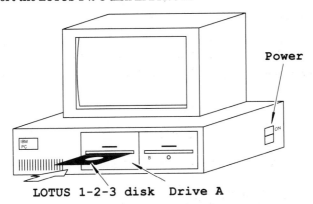

LOTUS 1-2-3 disk Drive A

Figure 1.2

7. Type *lotus*.

NOTE: Type "lotus" using either capital or lowercase letters.

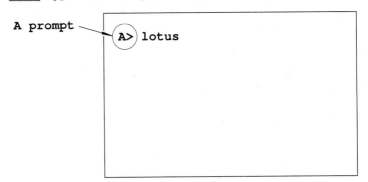

Figure 1.3

8. Press [↵] . This will load the LOTUS 1-2-3 program into the computer.

9. After the red light on Drive A disappears, press [↵] again.

10. Insert the ATHLETIC disk in Drive B.

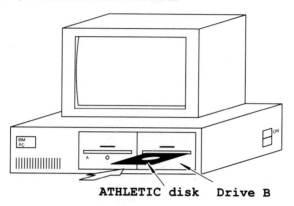

ATHLETIC disk   Drive B

Figure 1.4

11. Press [?/] . This will show you the main menu of the LOTUS 1-2-3 program.

12. Press [F] . This will show you the options in the file directory.

13. Press [R] . This will show you the files that are on the ATHLETIC disk.

14. Press ⏎ . The BUDGET file is now being loaded and the BUDGET structure will appear on the screen, as shown in Figure 1.5.

	A	B	C	D	E	F	G	H
1				MEN'S ATHLETIC BUDGET				
2	* * * *	* * * *	* * * *	* * * *	* * * *	* * * *	* * * *	* * * *
3	SPORT		VENDOR		ENCOMB	EXPEND	TOTAL EX	BALANCE
4								
5	BEGINNING BALANCE							30,000.00
6								
7								
8								
9								
10								
11								
12								
13								
14								
15								
16								

Figure 1.5

15. Proceed to the respective chapters if you want to add records, save records, change records, print records, or quit the program.

CHAPTER 2

*** ADD RECORDS ***

****** Follow steps 1—15 in "START UP" if the ATHLETIC ******
****** disk has not been loaded yet. ******

1. Using the arrow keys shown, move the cursor to column A. The
   column letter and row number of the cursor are indicated at the
   top of the screen, as shown in Figure 2.2.

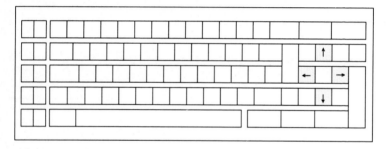

Figure 2.1

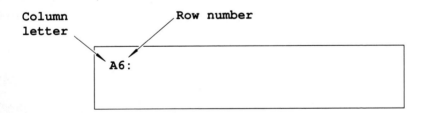

Figure 2.2

2. Using the arrow keys, move the cursor to the first empty cell in column A. This cell will be located in a row below the cell containing "BEGINNING BALANCE."

	A	B	C	D	E	F	G	H
1			MEN	'S ATHLETIC BUDGET				
2	★ ★ ★ ★	★ ★ ★ ★	★ ★ ★ ★	★ ★ ★ ★	★ ★ ★ ★	★ ★ ★ ★	★ ★ ★ ★	★ ★ ★ ★
3	SPORT		VENDOR		ENCOMB	EXPEND	TOTAL EX	BALANCE
4								
5	BEGINNING BALANCE							
6								
7								
8								
9								
10								
11								
12								
13								
14								
15								
16								

Figure 2.3

3. Type the sport to which the purchase will be charged in cell A.

	A	B	C	D	E	F	G	H
1			MEN	'S ATHLETIC BUDGET				
2	★ ★ ★	★ ★ ★ ★	★ ★ ★ ★	★ ★ ★	★ ★ ★ ★	★ ★ ★ ★	★ ★ ★ ★	★ ★ ★ ★
3	SPORT		VENDOR		ENCOMB	EXPEND	TOTAL EX	BALANCE
4								
5	BEGINNING BALANCE							
6	FOOTBALL							
7								
8								
9								
10								
11								
12								
13								
14								
15								
16								

Figure 2.4

4. Press → twice. This will shift the cursor to column C, which contains the vendor of the purchase.

5. Type the vendor of the purchase.

	A	B	C	D	E	F	G	H
1			MEN'S ATHLETIC BUDGET					
2	* * * *	* * * *	* * * *	* * * *	* * * *	* * * *	* * * *	* * * *
3	SPORT		VENDOR		ENCOMB	EXPEND	TOTAL EX	BALANCE
4								
5	BEGINNING BALANCE							
6	FOOTBALL		BADGER					
7								
8								
9								
10								
11								
12								
13								
14								
15								
16								

Figure 2.5

6. Press ⟶ twice. This will shift the cursor to column E, which contains the encumbered amount of the purchase.

7. Type the encumbered amount of the purchase.

   <u>NOTE</u>: This amount can be entered with or without decimals, as shown below.

   4000 or 4000.00

	A	B	C	D	E	F	G	H
1			MEN'S ATHLETIC BUDGET					
2	* * * *	* * * *	* * * *	* * * *	* * * *	* * * *	* * * *	* * * *
3	SPORT		VENDOR		ENCOMB	EXPEND	TOTAL EX	BALANCE
4								
5	BEGINNING BALANCE							
6	FOOTBALL		BADGER		1000.00			
7								
8								
9								
10								
11								
12								
13								
14								
15								
16								

Figure 2.6

8. Press [→] This will shift the cursor to column F, which contains the expenditure.

9. Type in the actual amount of the purchase.

   NOTE: This amount can be entered with or without decimals, as shown below.

      4000 or 4000.00

	A	B	C	D	E	F	G	H
1			MEN	'S ATHLETIC BUDGET				
2	* * * *	* * * *	* * * *	* * * *	* * * *	* * * *	* * * *	* * * *
3	SPORT		VENDOR		ENCOMB	EXPEND	TOTAL EX	BALANCE
4								
5	BEGINNING BALANCE							
6	FOOTBALL		BADGER		1000.00	1000.00		
7								
8								
9								
10								
11								
12								
13								
14								
15								
16								

Figure 2.7

10. Press [↵] .

45.6

* *

At this point, the entire record has been entered. The program will automatically compute the total expenditures to date and the balance to date. Actual output on the screen will appear as in Figure 2.8.

* *

	A	B	C	D	E	F	G	H
1			MEN'S ATHLETIC BUDGET					
2	* * *	* * * *	* * * *	* * * *	* * * *	* * * *	* * * *	* * * *
3	SPORT		VENDOR		ENCOMB	EXPEND	TOTAL EX	BALANCE
4								
5	BEGINNING BALANCE							30,000.00
6	FOOTBALL		BADGER		1000.00	1000.00	1000.00	29,000.00
7								
8								
9								
10								
11								
12								
13								
14								
15								
16								

Figure 2.8

11. Repeat steps 1–8 to add more records.

   NOTE: A maximum of 500 records can be entered in the BUDGET
   program.

12. Once you have entered the new records, proceed to Chapter 3,
   "SAVE RECORDS," to save these records in the BUDGET file.

CHAPTER 3

*** SAVE RECORDS ***

1. Press ⌜?/⌟ . The main menu will appear at the top of the screen.

2. Press ⌜F⌟ . This will show you what is in the file directory.

3. Press ⌜S⌟ . This will select the save option. "BUDGET" will be high-lighted by the cursor.

4. Press ⌜↵⌟ . This will save the new information in the BUDGET file.

5. Press ⌜R⌟ . This will replace the old information in the file with the new information.

6. Press ⌜Q⌟ . This will return you to the ready mode.

7. Refer to Chapters 2, 4, 5, or 6 to add more records, change records, print records, or quit the program.

CHAPTER 4

*** CHANGE RECORDS ***

******   Follow steps 1–15 in "Start Up" if the ATHLETIC   ******
******   disk has not been loaded yet.   ******

1. Using the four arrow keys, move the cursor to the cell containing the incorrect information.

   EXAMPLE: You found out the purchase listed in row 13 was bought from "BADGER," not "KING."

45.6

Move the cursor to cell C13, as shown in Figure 4.1.

	A	B	C	D	E	F	G	H
1			MEN	'S ATHLETIC BUDGET				
2	* * * *	* * * *	* * * *	* * * *	* * * *	* * * *	* * * *	* * * *
3	SPORT		VENDOR		ENCOMB	EXPEND	TOTAL EX	BALANCE
4								
5	BEGINNING BALANCE							30,000.00
6								
7								
8								
9								
10								
11								
12								
13	GOLF		KING		250.00	250.00		
14								
15								
16								

Figure 4.1

2. Type the correct information into the cell.

3. Press ⏎ . This will enter the new information, as shown in Figure 4.2.

	A	B	C	D	E	F	G	H
1			MEN	'S ATHLETIC BUDGET				
2	* * * *	* * * *	* * * *	* * * *	* * * *	* * * *	* * * *	* * * *
3	SPORT		VENDOR		ENCOMB	EXPEND	TOTAL EX	BALANCE
4								
5	BEGINNING BALANCE							30,000.00
6								
7								
8								
9								
10								
11								
12								
13	GOLF		BADGER		250.00	250.00		
14								
15								
16								

Figure 4.2

45.6

4. Repeat steps 1–3 to change other information.

5. Save the new information by referring to Chapter 3, "SAVE REC-
ORDS."

CHAPTER 5

*** PRINT RECORDS ***

****** Follow steps 1–15 in "START UP" if the ATHLETIC ******
****** disk has not been loaded yet.   ******

1. Press [?/] . The main menu will appear at the top of the screen.

2. Press [P] . This will show you what is in the print directory.

3. Press [P] . This will select the printer option.

4. Press [R] . This will select the range option.

5. Type in the letter of the first column, number of the first row, letter of the last column, and the number of the last row for the information you want to print.

EXAMPLE: You want to print the information contained in columns A through G and rows 1 through 10.

Type: A1 . . . . . G10

	A	B	C	D	E	F	G	H
1			MEN'S ATHLETIC BUDGET					
2	* * * *	* * * *	* * * *	* * * *	* * * *	* * * *	* * * *	* * * *
3	SPORT		VENDOR		ENCOMB	EXPEND	TOTAL EX	BALANCE
4								
5								
6	BEGINNING BALANCE							30,000.00
7	FOOTBALL		BADGER		1000.00	1000.00	1000.00	29,000.00
8	BASEBALL		GOPHER		1000.00	1500.00	2500.00	27,500.00
9	FOOTBALL		BADGER		100.00	100.00	2600.00	27,400.00
10	BASKETBALL		KING		750.00	750.00	3350.00	26,650.00
11	FOOTBALL		INDY		4000.00	4000.00	7350.00	22,650.00
12								
13								
14								
15								
16								

Figure 5.1

6. Press ⏎ . This will enter the print range.

7. Press ⬚G . The output will start printing. A typical printout is shown in Figure 5.2.

```
                        MEN'S ATHLETIC BUDGET
XXXXXXXXXXXXXXXXXXXXXXXXXXXXXXXXXXXXXXXXXXXXXXXXXXXXXXXXXXXXXXXXXXXXX
SPORT              VENDOR         ENCUMB     EXPEND    TOTAL     BALANCE
                                                      EXPEND

BEGINNING BALANCE                                               30000.00
FOOTBALL           BADGER         1000.00    1000.00   1000.00  29000.00
BASEBALL           GOPHER         1000.00    1500.00   2500.00  27500.00
FOOTBALL           BADGER          100.00     100.00   2600.00  27400.00
BASKETBALL         KING            750.00     750.00   3350.00  26650.00
FOOTBALL           INDY           4000.00    4000.00   7350.00  22650.00
```

Figure 5.2

8. After the print directory has returned to the top of the screen,

   press ⬚Q . This will return you to the ready mode.

9. Now you can add records, save records, change records, print more records, or quit the program by referring to Chapters 2 through 6.

CHAPTER 6

*** QUIT ***

<u>WARNING</u>: ANY RECORDS NOT SAVED BEFORE QUITTING WILL BE
LOST!

1. Press ⌈?/⌋ . The main menu will appear at the top of the screen.

2. Press ⌈Q⌋ . This will select the quit option.

3. Press ⌈Y⌋ . This will exit you from LOTUS 1-2-3.

4. Remove the LOTUS disk from Drive A, and the ATHLETIC disk from Drive B.

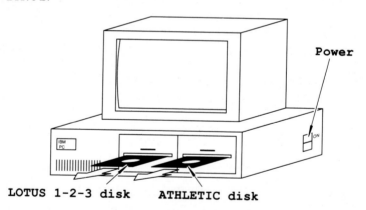

LOTUS 1-2-3 disk   ATHLETIC disk

Figure 6.1

5. Turn the power off.

GLOSSARY OF USAGE

This glossary lists the most commonly misused or misunderstood words that you may encounter as a business or technical writer. Current usage is a factor in the explanations.

The terms *technical writing* and *colloquial* English are employed throughout this glossary and need to be clarified. *Technical writing* refers to all formal or informal precise material written for an educated audience; it must be correct to the last detail. Of course, all writers should meet these standards, but the technical writer must be even more exact. *Colloquial* English refers to spoken words that are sometimes acceptable in everyday speech but are not acceptable in written material.

Two other terms are employed throughout the glossary: *nonstandard* English and *jargon*. *Nonstandard* English is usage unacceptable in any situation, written or spoken. *Jargon*, as used in this glossary, is language particular to the world of business and industry.

The entries in this glossary are meant to serve as guidelines for you, not as absolute rules. Although the glossary can offer helpful advice when you have to make a decision about a word or phrase, the final choices are always up to you.

a, an

A is an indefinite article that you use before a noun, an adjective, or an adverb that begins with a consonant sound.

EXAMPLES

a cylinder, a histogram, a university

An is an indefinite article that you use before a noun, an adjective, or an adverb that begins with a vowel sound.

EXAMPLES

an arc, an herb, an umbrella

absolutely

Absolutely is an adverb that means *entirely, totally,* or *without a question* and does not replace the intensifier *very.*

weak	improved
The chairman told the board it was *absolutely* wrong in its decision.	The chairman told the board it was very wrong in its decision.

accept, except

Accept is a verb that means *to receive willingly. Except* is a preposition that means *other than.*

EXAMPLES

The candidate *accepted* the nomination.
Everyone was promoted *except* Stephen.

access, excess

Access is *admittance* or *liberty. Excess* is *the state of going beyond normal limits.*

EXAMPLES

All employees have *access* to the files on the third floor.
The deficit was in *excess* of $12,000.

accumulative, cumulative

Accumulative and *cumulative* are synonyms for *adding wealth, ability, or power, usually over a period of time.* Normally, use *accumulative* in reference to accumulative wealth and use *cumulative* in reference to cumulative power or ability.

weak	improved
The Johnson's *cumulative* wealth was not known until after their deaths.	The *cumulative* effects of the radiation leak will not be known for many years.

activate, actuate

Both *activate* and *actuate* are synonyms for *making active*. *Activate* refers to making most things active. *Actuate* refers to making processes, such as mechanical and chemical ones, active.

EXAMPLES

The company *activated* Plan C.
The switch *actuates* the monitor.

actually

Actually is an adverb that means *really* or *in reality*. It is useful in speech to emphasize a word; however, because the word does not add anything to a sentence, avoid using *actually* in technical writing.

AD or A.D.

AD means *Anno Domini, in the year of our Lord;* therefore, it is redundant to write "in the year *AD* 1602." Always place *AD* before the year, not after it.

EXAMPLE

He was born in *AD* 1952.

ad, advertisement

Ad is a shortened form of *advertisement*. Always use *advertisement* in technical writing.

adapt, adept, adopt

Adapt is a verb that means *to make something fit* or *to adjust to a new situation*. *Adept* is an adjective that means *to be very skilled*. *Adopt* is a verb that means *to take or use something voluntarily that is not one's originally*.

EXAMPLES

Marshall can *adapt* to new situations more quickly than anyone else in the firm.

He is *adept* at making changes in the computer.

The quality-control department has decided to *adopt* the coffee-break policy used in the publications department.

advice, advise

Because these two words sound similar, they are often confused. *Advice* is a noun that refers to the suggestion that is given to someone. *Advise* is a verb that refers to giving advice.

EXAMPLES

The doctor gave the patient some *advice* about his weight.

The doctor *advised* the patient about his weight.

advise, inform

Advise is a verb that means *to give suggestions. Inform* is a verb that means *to communicate information.*

EXAMPLES

I *advise* you to enter our profit-sharing plan.

I *inform* you that we have a profit-sharing plan.

affect, effect

Affect is a verb that means *to change* or *to influence. Effect* is a verb or a noun. As a noun it means the *result* or *outcome.* As a verb it means *to bring about a change.*

EXAMPLES

The decline in the stock market *affected* the company's profits. [verb]

The *effect* of the stock market's decline was seen in the company's profits. [noun]

The chair of the board *effected* a change in the investment policies of the company. [verb]

afraid, frightened, scared   See **frightened, scared, afraid.**

again, back
Do not use *again* and *back* with words beginning with *re-* that already imply *back* or *again*.

EXAMPLES

recount, rework, rehabilitate, restate

weak	**improved**
He recounted the incident *again*.	He recounted the incident.
Alan quickly rebounded *back*.	Alan quickly rebounded.

aggravate
Aggravate is a verb that means *to make worse*. Do not use *aggravate* to replace *annoy* or *irritate*.

weak	**improved**
The noise *aggravated* my boss.	The noise irritated (or annoyed) my boss.
	but
	The noise *aggravated* my boss's headache.

a half a, half a, half an
Half a and *half an* are correct uses of half; however, *a half a* is not. Substitute either *half a* or *a half* for *a half a*.

weak	**improved**
A *half a* loaf is better than none.	*Half a* loaf is better than none.

all around, all-around, all-round
All around is a term used to designate that something continues in a given space, such as "*all around* the cobbler's bench." *All-around* or *all-round* refers to something or someone who is versatile or accomplished, such as "the *all-around* (or *all-round*) athlete."

allude, refer
To *allude* to something means *to suggest* it or *indirectly infer* it. To *refer* to something means *to make direct mention* of it.

EXAMPLES

My boss *alluded* to my absences when he spoke of employees who take sick-leave vacations.

My boss *referred* to my absences when he told me I had used 15 days of sick leave last year.

allusion, illusion

An *allusion* is *a suggestion of indirect inference about someone or something.* An *illusion* is *a false notion or a mistaken idea about someone or something.*

EXAMPLES

His *allusion* to "things undone" left the rest of us wondering if the company was going to make some changes.

Martha is under the *illusion* that she will be promoted to assistant manager this spring.

a lot, alot

Alot is not a word. *A lot* is the correct spelling. Avoid the use of *a lot* in technical writing because it is informal and often awkward. Use a term such as *much* instead.

weak	improved
We have *a lot* to do next week.	We have much to do next week.

already, all ready

Already means something has happened before the present time. *All ready* means something is completed or that all are ready.

EXAMPLES

We *already* have the proposed figures for next year.

We are *all ready* to give our presentation.

alright, all right

Do not use *alright* because it is an incorrect spelling of *all right. All right* means that everything is correct, not that something is acceptable.

weak	**improved**
Her answers on the test were *alright*.	Her answers on the test were *all right*.

also
Do not use *also* in place of *and*.

weak	**improved**
The bridge club held a charity tournament, *also* a bazaar.	The bridge club held a charity tournament and a bazaar.

altogether, all together
Altogether means *entirely*. *All together* means *all are in one place or are acting together*.

EXAMPLES

Our figures for the assembly were *altogether* wrong. It isn't often that we are *all together* in the office.

AM, PM or am, pm
When *AM* and *PM* are used as nouns, they are redundant. It is also redundant to modify *AM* or *PM* with prepositional phrases.

EXAMPLES

He arrived at 9 *in the PM*. (Omit *in the*.)
He arrived at 9 *PM in the night*. (Omit *in the night*.)

among, between
Among and *between* are two prepositions that are often confused. *Among* is used with more than two objects and means *surrounded by*. *Between* is used with relationships between two objects.

EXAMPLES

He walked cautiously *among* the noisy party-goers.
She had to choose *between* two well-paying, satisfying jobs.

amount of, number of
Amount of refers to the total number and is followed by a singular noun. *Number of* refers to a sum and is followed by a plural noun.

EXAMPLES

> She received a large *amount* of money for the painting.
> She received a *number of* new commissions after her first painting was exhibited.

and etc.
In Latin, *etc.* is an abbreviation for *et cetera* meaning "and so forth." *Et* is Latin for *and;* therefore, if you use *and* with *etc.* you are being redundant.

and/or
And/or is acceptable in technical writing, although some writers still find it awkward and tend to avoid it. Whether to use it or not is a matter of personal preference.

angry, mad   See **mad, angry.**

a number, the number
Use *a number* as a plural noun. Use *the number* as a singular noun.

EXAMPLES

> *A number* of businesses *were* happy about the construction of the new highway.
> *The number* of businesses in Freedom Center *is* growing smaller every year.

anti-, ante-
Ante- is a prefix that means *before*. *Anti-* is a prefix that means *against*.

EXAMPLES

> antecedent, anterior, anteroom
> anticlimactic, antimagnetic, antipathy

anxious, eager
To be *anxious* is *to be filled with anxiety or worry*. To be *eager* is *to want or expect something*.

EXAMPLES

The *anxious* man waited for news about the accident.
The *eager* man waited for news about the contest he had entered.

anyone, any one; everyone, every one

Anyone used as one word refers to *any person. Any one* used as two words refers to one member of a group. *Everyone* used as one word refers to *all people. Every one* used as two words refers to each member of a group.

EXAMPLES

Anyone can attend the conference.
Any one of you can attend the conference.
Everyone can attend the conference.
Every one of you can attend the conference.

appendixes, appendices

Appendixes is the non-Latin plural of appendix. *Appendices* is the Latin plural of appendix. Either spelling is correct.

as   See like, as.

as far as

As far as is an unacceptable substitute for *as for*.

weak	improved
As far as earning a living, Gretchen has never tried.	As for earning a living, Gretchen has never tried.

as per

As per is an unacceptable substitute for *in accordance with*.

weak	improved
As per your instructions, I completed the report.	In accordance with your instructions, I completed the report.

aspect

Aspect means *a view from a vantage point* and not *consideration*.

weak	improved
The report has five *aspects* to deal with.	The report has five considerations to be made.

assure, insure, ensure   See **insure, ensure, assure.**

augment, supplement
Augment means *to increase in size or effect. Supplement* means *to add something, often to make up a deficiency.*

EXAMPLE

> The sales department *augmented* their year-end report with a 200-page *supplement.*

author
Do not use *author* as a verb.

weak	improved
He *authored* our club's constitution.	He wrote our club's constitution.

average, mean, median
The *average* refers to the *mean* or *median* and is determined by using one of two different methods that should not be used in place of one another. The *mean* is found by adding all the numbers in a given series and dividing by the number of items. For example, if five numbers are added together, the sum is divided by five. The *median* is the middle number in a series of numbers that are placed in order, beginning with the lowest or highest quantity. For example, in the series 1, 2, 3, 4, 5, *3* is the median.

awful
Awful is conversational and is often overused for *disgusting, inferior,* and other pejorative words. Its formal meaning, *awesome or inspiring fear,* is rarely used. Because of misuse, avoid using *awful* in technical writing.

awhile, a while
Awhile is an adverb that means *for a short time. A while* is a noun that means *a short time. Awhile* cannot be used with a preposition.

EXAMPLES

We waited *awhile* before we boarded the plane.
We waited for *a while* before we boarded the plane.

back, again   See **again, back.**

back of
Back of is a colloquial term for *behind* that you should avoid using in technical writing.

weak	**improved**
He was *back of* the house.	He was behind the house.

backward/backwards
Backward is an adjective used in a construction such as "The *backward* child attended the public school." *Backward* or *backwards* is an adverb used in a construction such as "She spelled the word *backward* (or *backwards*)."

bad, badly
Badly is an adverb often misused with the linking verbs *to feel* or *to look* when *bad,* an adjective, is substituted for *ill.* (Also see **poorly.**) *Badly* is an adverb that is preferred after most verbs.

weak	**improved**
Harry feels *badly* today.	Harry feels *bad* today.
Tim did *bad* on the test.	Tim did *badly* on the test.

balance, remainder
Balance and *remainder* are often confused, both when used as bookkeeping terms and in mathematics. *Balance* refers to what is left in an account while *remainder* refers to what is left over after an adjustment.

EXAMPLES

After the Christmas party, the *balance* in the party fund was 12 cents.
After much multiplying and dividing, John had a *remainder* of 23, which was not the right answer to the question.

In other uses, *balance* and *remainder* are sometimes synonymous.

EXAMPLES

The *balance* of the speakers were from three companies.
The *remainder* of the speakers were from three companies.

being as, being that
Being as and *being that* are nonstandard usages for *since* or *because*.
Do not use either term in technical writing.

weak	improved
Being as (or *being that*) we had three members of the winning bowling team in our department, we had a celebration.	Because we had three members of the winning bowling team in our department, we had a celebration.

beside, besides
Beside is a preposition meaning *next to*. *Besides* is an adverb meaning *in addition to*.

EXAMPLES

I sat down *beside* Jim.
Besides all his other faults, he was perpetually tardy.

be sure and, be sure to
Be sure and is unacceptable usage for *be sure to*. Use *be sure to* in technical writing.

weak	improved
Be sure and remember to attend the lecture on Thursday.	*Be sure to* remember to attend the lecture on Thursday.

better, had better
Better is confined to spoken English. In technical writing, use *had better* as the preferred usage.

weak	improved
The surveying crew *better* return by 6 PM.	The surveying crew *had better* return by 6 PM.

between   See **among, between.**

between you and me
Between is a preposition. Pronouns are objects of prepositions. Therefore, the correct phrase is *between you and me* and not *between you and I*, which is frequently employed.

bi, semi
When used to refer to time, *bi* means *two* or *every two*. *Semi* means *half*.

EXAMPLES

We receive the marketing report *bimonthly* (once every two months).
We receive the marketing report *semimonthly* (once every two weeks).

biannual, biennial
Biannual refers to twice a year, while *biennial* refers to once every two years.

bias, prejudice
Bias and *prejudice* can be used synonymously to indicate a negative preconceived view, but *bias* also can be positive while *prejudice* is always negative.

EXAMPLES

His *prejudice* has led to his hatred of many groups.
His *bias* toward helping the poor is responsible for his heading this year's food drive.

big, large, great
Big refers to bulk or weight. *Large* refers to dimensions or capacity. *Great* refers to importance.

EXAMPLES

The *big* man wore a size 48 suit.
The *large* warehouse is on Water Street.
The *great* leader was interviewed on television.

borrow off, borrow from
Borrow off is used in conversation. *Borrow from* is the correct usage for technical writing.

weak	improved
This is the book I *borrowed off* Gerald.	This is the book I *borrowed from* Gerald.

bottom line
Bottom line has been overused. Use *outcome* or *final result* as more acceptable substitutes in technical writing.

brake, break
To *brake* is *to come to a stop*. To *break* is *to smash something* or *to divide something into pieces or parts of the whole*.

EXAMPLES

He *braked* at the stoplight.
She didn't mean *to break* her good china teapot.

broke
Broke is conversational for *to have no money*. *Broke* also is substandard usage for *broken*.

weak	improved
She is *broke* this week.	She has no money this week.
The large white urn was *broke*.	The large white urn was *broken*.

but what, that
But what is a conversational expression that is meaningless because it serves no grammatical function. Substitute *that* for *but what* in technical writing.

weak	improved
I have no doubt *but what* I will be asked to write the report.	I have no doubt *that* I will be asked to write the report.

can, may
When permission is sought, these two words are interchangeable. In formal writing a distinction is made between the two words. *Can* refers to ability and *may* refers to permission. Decide whether to use the *can/may* distinction based on the formality of your writing.

EXAMPLES

> *Can* the typists use Word Perfect? (Are they capable?)
> *May* the typists use Word Perfect? (Do they have permission?)

cannot, can not
Write *cannot* as one word.

cannot help but
Cannot help but is a double negative (two negatives used together when only one is necessary) that you should not use in any writing.

can't hardly, can't scarcely
Can't hardly and *can't scarcely* are double negatives (two negatives used together when only one is necessary). Use *can hardly* or *can scarcely* instead.

weak	improved
The child *can't hardly* wait for Christmas.	The child *can hardly* wait for Christmas.

canvas, canvass
Canvas is a noun that describes a heavy woven fabric often used in tents. *Canvass* is a verb that refers to the act of soliciting votes or opinions.

EXAMPLES

> The *canvas* tent flapped noisily in the wind.
> The *canvass* of the local voters showed that most of them were opposed to the referendum.

capital, capitol
A *capital* is *the center of government for a state or nation.* A *capitol* is *the building that houses the center of government.*

EXAMPLES

> The *capital* of Wisconsin is Madison.
> The *capitol* is very impressive when it is lit up at night.

censor, censure
To *censor* is *to examine objectionable material and determine that it should be deleted.* To *censure* is *to reprimand* or *to condemn.*

EXAMPLES

> The citizens committee met to determine which books to *censor* in the high school library.
> The PTA voted to *censure* the citizens committee for the books they had *censored* in the school library.

center, middle
The *center* is the point around which everything revolves. The *middle* is a space.

EXAMPLES

> She was the *center* of attraction wherever she went.
> She found herself in the *middle* of a controversy.

cite, site, sight
Cite is a verb that refers to quoting an authority. *Site* is a noun that refers to a place where something is located. *Sight* is a noun or a verb that refers to vision or the ability to see.

EXAMPLES

> He *cited* Rutherby as an authority on bee keeping.
> The construction *site* was shut down because of the local strike.
> He *sighted* the woods before I did. [verb]
> His *sight* is slowly diminishing. [noun]

class, classify
Class is not a substitute for *classify.* Do not use it as such in technical writing.

weak	improved
He *classed* the information.	He *classified* the information.

compare to, compare with
To compare to is *to regard as similar. To compare with* is *to compare or contrast.*

EXAMPLES

> Every time Senator Clogg gave a speech, he *compared* the Senate *to* a football game.
> Kevin *compared* the old blueprint machine *with* the new CAD/CAM program.

complement, compliment
Complement means *completing the whole* or *supplying a need. Compliment* means *praise.*

EXAMPLES

> Sharon wore red shoes to *complement* her outfit.
> Sharon received several *compliments* when she wore her red shoes with her black and red outfit.

compose, comprise
Compose is a verb that means *to create* or *to be part of a whole. Comprise* is a verb that means *to include.*

EXAMPLES

> Three office managers *compose* the memos for all meetings.
> The secretarial pool is *composed of* twelve members.
> The twelve secretaries *comprise* the secretarial pool.

concept, conception, idea
A *concept* is *an abstract notion.* A *conception* is *a mental picture or understanding that can be wrong. Idea* can mean either a *concept* or a *conception* and would fit in either of the above sentences.

EXAMPLES

> The *concept* is one that most members of the club accept.
> Her *conception* of how to operate the copy machine has led to many repair bills.

conscious, conscience
Conscious is an adjective meaning *aware* or *able to feel and think. Conscience* is a noun meaning *the sense of right and wrong.*

EXAMPLES

She is *conscious* that she made the error, but she won't admit that she
made it.

Her *conscience* bothers her because she made the error and won't
admit it.

consequently, subsequently   See **subsequently, consequently.**

contemptible, contemptuous
Contemptible means *deserving of contempt. Contemptuous* means
showing contempt.

EXAMPLE

The *contemptible* thief was *contemptuous* of the police.

continual, continuous
Continual means *happening frequently. Continuous* means *happening without interruption.*

EXAMPLES

The *continual* turnover in managers was discouraging to the rest of
the staff.

The *continuous* noise from the printers was often distracting to the
graphic artists.

could of
Could of is nonstandard usage for *could have.* See **of** for a more
complete discussion.

couple, couple of
These terms are colloquial for *two.* Use *two* in technical writing,
where precision is necessary.

credible, creditable
Credible is *believable. Creditable* is praiseworthy.

EXAMPLES

His story is *credible.*
His work is *creditable.*

criterion, criteria, criterions
Criterion is singular for *a standard of judging*. Both *criteria and criterions* are plural forms of *criterion*, but *criteria* is generally preferred.

cumulative   See **accumulative, cumulative.**

data
Data is plural for *datum;* however, many writers consider *data* to be a collective noun and treat it as singular. It has become a matter of individual preference whether to treat the term as plural or singular, although some writers treat *data* as plural when referring to *facts* and as singular when referring to *information.* You must decide how to treat *data* in your technical writing.

deduce, deduct
To *deduce* is *to infer*. To *deduct* is *to take away*.

EXAMPLES

Sherlock Holmes *deduced* that the man recently had been for a walk on the moor.
He has to *deduct* what he owes on his loan from his paycheck.

defective, deficient
To be *defective* is *to be faulty*. To be *deficient* is *to be lacking in something*.

EXAMPLES

The new copy machine was *defective*.
Her tests showed she was *deficient* in her potassium level.

different from, different than
Different from is more formal than *different than* and you should use it in most technical writing. Use *different than* only if it is followed by a clause.

EXAMPLES

The Apple 2e is *different from* the Apple MacIntosh.
The symposium on retailing was *different from* what I had expected.

or

The symposium on retailing was *different than* I had expected.

differ from, differ with
Differ from means *to be not like something else. Differ with* means *to be in disagreement with someone else.*

EXAMPLES

Barry *differs from* his twin brother Larry.
Even though Barry and Larry are twins, they frequently *differ with* one another.

disinterested, uninterested
Disinterested means *impartial. Uninterested* means *indifferent.*

EXAMPLES

The *disinterested* referee made an unpopular decision with the home-team fans.
The *uninterested* bystander ignored the band as it marched by.

doubtlessly
Adding -ly to the adverb *doubtless* creates a redundancy. Avoid *doubtlessly* in formal writing.

due to, because of
Although *due to* is gaining more acceptance as a replacement for *because of,* many experts do not agree. You probably should not replace *because of* with *due to,* although the final decision is yours.

weak	improved
Due to a high weather front, we had a great deal of rain.	*Because of* a high weather front, we had a great deal of rain.

each and every
Although *each and every* is frequently employed in conversation, avoid using it in technical writing because it is repetitious. *Each* or *every* by itself is preferable.

eager, anxious See **anxious, eager.**

economic, economical

Economic is an adjective that refers to production, distribution, and consumption of wealth. *Economical* is an adjective that refers to thriftiness.

EXAMPLES

The *economic* value of the new factory in the community will be felt by next year.

She decided to purchase an *economical* compact car.

effect, affect   See **affect, effect.**

electric, electrical

Electric is *anything that conveys or is charged with electricity. Electrical* is *anything that pertains to electricity.*

EXAMPLES

They bought an *electric* ice-cream maker for Christmas.
The *electrical* engineer works for his father.

emigrate from, immigrate to

To emigrate from is *to leave one place to settle in another. To immigrate to* is *to settle in a place after leaving another.*

EXAMPLES

My grandparents *emigrated from* Sweden.
My grandparents *immigrated to* the United States.

eminent, imminent, immanent

An *eminent* person is one who is *distinguished.* An *imminent* event is one that is *about to happen.* An *immanent* being is one who is *inherent* in the universe.

EXAMPLES

The *eminent* doctor spoke at the dinner.
The baby's arrival is *imminent.*
Many people believe in an *immanent* being.

enable

To enable is a verb that means *to make able.* Do not use *to enable* in place of *to allow.*

weak	improved
The legislature *enabled* the teachers to receive a raise.	The passage of the bill *enabled* the state to give the teachers a raise.

ensure, assure, insure   See **insure, ensure, assure.**

enthuse

Enthuse is colloquial for *to show enthusiasm.* Do not use *enthuse* in technical writing.

especially

Especially means *outstandingly. Specially* means specifically.

EXAMPLES

   The concert was *especially* enjoyable.
   He *specially* wanted me to notice his new jacket.

etc.

Etc. is used in conversational writing in place of *and so forth.* Avoid its usage in technical writing when precision is important.

ever so often, every so often

Ever so often means *frequently. Every so often* means *occasionally.*

everyone, every one   See **anyone, any one.**

exceedingly, excessively

You must be very careful not to confuse these two terms. *Exceedingly* means very much; *excessively* means too much.

EXAMPLES

   The work went *exceedingly* well.
   The James family spends money *excessively.*

excess, access   See **access, excess.**

excuse, pardon
To excuse is to overlook a fault. *To pardon* is to forgive an offense.

explicit, implicit, tacit
Something *explicit* is stated directly while something *implicit* is suggested but never stated directly. *Tacit* means *silent.*

EXAMPLES

The *explicit* statement was that all personnel were required to take a TB test.
Although my boss never told me directly that I was late too frequently, I understood his *implicit* meaning.
The two companies made a *tacit* agreement to merge in three years.

fantastic
Fantastic is an overused and colloquial term for *remarkable* that you should avoid in technical writing.

farther, further
Although these two terms are used interchangeably, they do have slightly different meanings. *Farther* refers to distance while *further* refers to time or to an addition.

EXAMPLES

Paula can run *farther* than Annette.
To *further* clarify his point, Anthony pounded on the podium.

fewer, less
Normally use *fewer* with objects that can be counted, and *less* with items that can be measured.

EXAMPLES

fewer people, *fewer* farms, *fewer* bushels of apples
less grain, *less* material, *less* work

figuratively, literally
Although *figuratively* means *not literally,* it is often confused with *literally. Literally* means *actually.*

EXAMPLES

> He was speaking *figuratively* when he said that he felt like a million dollars after he read the sales report.
>
> He was speaking *literally* when he said that the company sales had almost doubled in the last three years.

first, firstly
First is an adverb that does not need the suffix *-ly* added to it to make it an adverb. Use *first*, not *firstly*, in technical writing.

fixing to
Fixing to is colloquial usage for *getting ready to*, or *preparing*.

weak	improved
He was *fixing to* retire.	He was getting ready to retire.

flammable, inflammable, nonflammable
Both *flammable* and *inflammable* mean *easily set on fire*. *Flammable* is used in most industries because *in-* means *not;* thus the term *inflammable* causes confusion. *Nonflammable* means *not easily set on fire*.

flunk
Flunk is a colloquial term for *fail*. Do not use *flunk* in technical writing.

weak	improved
The soil samples *flunked* the perk test.	The soil samples *failed* the perk test.

forceful, forcible
Both *forceful* and *forcible* are adjectives that mean *full of force*. Use *forceful* to designate persuasive ability, such as "a forceful speaker." Use *forcible* to designate physical force, such as "a forcible entry."

former, latter
Use *former* and *latter* only when a reference is to two people or items. If more than two items are referred to, use *first* and *last*.

frightened, scared, afraid
Both *frightened* and *scared* suggest that something immediately causes that reaction. *Afraid* refers to a persistent fear or regret.

EXAMPLES

The children were *frightened* (or *scared*) of the thunder and lightning.
Janice was *afraid* her car wouldn't last through the winter.

go, goes
Do not use *go* or *goes* as a substitute for *say* or *says* in technical writing.

weak	improved
Every time Sam sees me, he goes "Hi, Betty!" even though my name is Frances.	Every time Sam sees me, he says "Hi, Betty!" even though my name is Frances.

good, well
Do not confuse the adjective *good* with the adverb *well*.

weak	improved
The student did good on his exam.	The student did well on his exam.

great
In technical writing use *great* to indicate the importance of a person, place, or idea, not to mean *clever*, *good*, or the like.

weak	improved
He is a great worker.	He is a good worker.
She is a good leader.	She is a great leader.

See **big, large, great** for further discussion of the term.

guy
Guy is a colloquial term for *person*.

had of, had have
Both *had of* and *had have* are nonstandard usage for *had*. Never use either one in technical writing.

half a, a half, a half a   See **a half a, half a, half an.**

hanged, hung
Hung is the past participle of the verb *to hang* when *hang* is used to mean hanging something, as in "She *hung* the clothes on the line." *Hung* is not the past participle of *to hang* when *hang* is used to mean to execute someone. *Hanged* is then the correct usage, as in "The traitor was *hanged* at sunrise."

healthy, healthful
Healthy means *to have good health. Healthful* means *to promote good health.*

EXAMPLES

> *healthy* children, *healthy* employees
> *healthful* meals, *healthful* exercise

herein, herewith
Both *herein* and *herewith* are business jargon that you should avoid in technical writing.

hisself
Hisself is not a word. Use *himself.*

hopefully
Hopefully is a controversial term used to replace *I hope* or *it is hoped.* Some authorities accept *hopefully;* others do not. It is your decision whether or not to use it.

how
How is ambiguous when it is used to replace *that.* Avoid using this construction in technical writing.

weak	improved
He told her *how* he was going fishing on Saturday.	He told her that he was going fishing on Saturday.

how come
How come is a conversational use of *why* that you should avoid in technical writing.

weak	improved
How come you are going to the rally rather than the dance?	Why are you going to the rally rather than the dance?

idea, concept, conception   See **concept, conception, idea.**

illusion   See **allusion, illusion.**

immigrate to   See **emigrate from, immigrate to.**

imminent, immanent, eminent   See **eminent, imminent, immanent.**

impact
Do not use *impact* as a synonym for *effect*. *Impact* means *to collide* or *to force together*, not *to influence*.

EXAMPLES

   The *impact* of the crash totalled both automobiles.
   The *effect* of the advertising campaign was to boost sales by 12 percent.

implicit, explicit, tacit   See **explicit, implicit, tacit.**

imply, infer
Imply means *to suggest*. *Infer* means *to conclude based on evidence*.

EXAMPLES

   He *implied* his work was superior to mine.
   I *inferred* that he was jealous of my promotion.

include
Use *include* before a listing when the listing is incomplete.

EXAMPLE

   Our order *includes* pens, legal pads, and pencils.

incredible, incredulous
Incredible is an adjective that means *unbelievable*. *Incredulous* is an adjective that means *not able to believe*. Statements and events are *incredible*, while people are *incredulous*.

EXAMPLES

Hans was fired for telling an *incredible* story about his two weeks' absence.

Because Jon believes almost anything he is told, most people in the office laugh about how *incredulous* he is.

indiscreet, indiscrete

Indiscreet means *lacking prudence. Indiscrete* means *not separated into distinct parts.*

EXAMPLES

Her conduct at the party was *indiscreet.*

In the fog, the shapes of the houses merged into one *indiscrete* mass.

inflammable   See **flammable, inflammable, nonflammable.**

inform, advise   See **advise, inform.**

ingenious, ingenuous

Ingenious means *clever. Ingenuous* means *open or frank.*

EXAMPLES

The *ingenious* invention was demonstrated at the convention.

Shelley's *ingenuous* attitude disarmed her opponent.

input

Input is a computer term that you should not use to mean *a voice in.*

weak	improved
The employees have no *input* in company policy.	The employees have no voice in company policy.

in regards to, with regards to

In regards to and *with regards to* are unacceptable forms of *in regard to* and *with regard to* that you should not use in technical writing.

insure, ensure, assure

Insure and *ensure* mean *to make secure from harm* and are widely used in the business world when referring to risk and the guarantee of freedom from it, such as in "The company insured against a loss" or "They bought the policy to ensure protection." *Assure* has the same meaning but is used with persons, as in "Leonard assures me that the investment is a sound one."

interface
Once used primarily as a sewing term, *interface* is now used widely in business and industry when referring to any surface that provides a common link between two objects.

EXAMPLE

The two computer systems have been *interfaced.*

irregardless, regardless
Irregardless is not a word. Do not substitute *irregardless* for the correct word, *regardless.*

its, it's
Its is the possessive form of *it.* *It's* is the contraction for *it is.*

EXAMPLES

its meaning, *its* flavor
it's fun, *it's* time

judicial, judicious
Judicial refers to the legal system: judges, law courts, or their functions. *Judicious* means *wise.*

EXAMPLES

The *judicial* system in the United States is often criticized for leniency.
My boss made a *judicious* decision when he decided to accept the Waterbury account.

kind of, sort of
Kind of and *sort of* are colloquial expressions of *a vague degree.*

weak	improved
I felt *kind of* sorry for her.	I felt sorry for her.

large, great, big   See **big, large, great.**

latter, former   See **former, latter.**

leave, let
Leave is a verb that you should never use in place of the verb *to let.*

weak	**improved**
Leave her do it herself.	*Let* her do it herself.

less, fewer   See **fewer, less.**

libel, liable, likely
Libel means *a false or malicious written statement about a person or a group. Liable* means *legally responsible for.*

EXAMPLES

> When our local newspaper accused our mayor of accepting bribes, he sued the paper for *libel* and won.
> Technical writers are held *liable* for mistakes they make in their manuals.

Likely means *probable* and should never be confused with *liable.*

weak	**improved**
He is *liable* to be caught.	It is *likely* that he will be caught.

like, as
Like is a preposition and *as* is a conjunction. The two words are frequently confused. Use *like* with nouns or pronouns that are not followed by a verb. Use *as* before clauses.

EXAMPLES

> The man acted *like* a maniac.
> She acted *as* we had predicted she would.

literally   See **figuratively, literally.**

lose, loose
Lose is a verb; *loose* is usually an adjective.

EXAMPLE

> She did not need to *lose* any more weight, for her clothes were already too *loose* on her.

mad, angry
Mad means *insane*. *Angry* is normally the term you want to use.

weak	**improved**
She is *mad* at me.	She is *angry* with me.

may, can   See **can, may.**

may be, maybe
May be is a verb phrase; *maybe* is an adverb.

EXAMPLES

> He *may be* (ought to be) going.
> *Maybe* (perhaps) he will go.

me and
Me and is usually nonstandard usage. Avoid it.

weak	**improved**
Me and my boss went to the symposium.	My boss and I went to the symposium.

media, medium
Media is plural for *medium* and is used with a plural verb. Unlike *data*, it is never a collective noun.

EXAMPLES

> Television is the most popular *medium.*
> *The media* are always blamed for over-dramatizing events.

middle, center   See **center, middle.**

mighty
Mighty is conversational for *very,* such as in *mighty fine.* Do not employ it in technical writing.

mix, mixture

Mix is a verb; *mixture* is a noun. Do not substitute *mix* for *mixture* in technical writing.

weak	improved
The *mix* was Rob's discovery.	The *mixture* was Rob's discovery.

morale, moral

Morale is a noun that refers to mood. *Moral* is chiefly an adjective that refers to the discernment of goodness and badness.

EXAMPLES

> The team *morale* was very low.
> The *moral* judgments people make can be quite surprising at times.

most, almost

Do not use *most* in place of *almost* in technical writing.

weak	improved
He was *most* finished.	He was *almost* finished.

Ms

Ms is a method of addressing a woman that does not depend on her marital status. It is widely used, although some women still object to its usage.

myself

Myself is not an acceptable replacement for the subjective *I* or for the objective *me*.

weak	improved
Harold and *myself* went to the opera.	Harold and *I* went to the opera.
He gave the photographs to Riley and *myself*.	He gave the photographs to Riley and *me*.

needless to say

Needless to say is a redundant business expression that you do not need to use in technical writing.

neo-
Neo- is a Greek prefix that means *new*. Always hyphenate *neo-* when it is used with a proper noun or with a word beginning with a vowel. When *neo-* is used with other nouns, do not hyphenate it.

EXAMPLES

Neo-Catholic, Neo-Scholasticism, neo-orthodoxy
neolithic, neoplastic, neologism

no doubt but
No doubt but is a wordy phrase that you should replace with *no doubt* in technical writing.

weak	improved
There is *no doubt but* she did it.	There is no doubt she did it.

no . . . nor
No . . . nor is not used in place of *no . . . or* in compound phrases.

weak	improved
They had *no* paper *nor* pens to take notes.	They had *no* paper *or* pens to take notes.

nonflammable   See **flammable, inflammable, nonflammable.**

not . . . nothing
When *not* is combined with *nothing* in a sentence, it normally produces a double negative.

weak	improved
He did *not* do *nothing* to deserve being fired.	He did *nothing* to deserve being fired.

noted, notorious
A *noted* person is *one who has attained fame or recognition for doing something desirable. A notorious person is one who has a negative reputation for doing something undesirable.*

EXAMPLES

The *noted* scientist gave the after-dinner speech.
He was a *notorious* criminal.

nowhere near

Nowhere near is a conversational expression that has no real meaning in a sentence. Do not use the term in technical writing.

weak	improved
We have *nowhere near* enough teachers to offer small classes.	We do not have nearly enough teachers to offer small classes.

number of, amount of   See **amount of, number of.**

observance, observation

An observance is *a customary act, rite, or duty. An observation* is *the act of seeing and recording or noting something.* Do not confuse the terms.

EXAMPLES

    Memorial Day is a yearly *observance.*
    By my *observation,* Christmas decorations appear in almost all stores
        and on lampposts two weeks before Thanksgiving Day.

occur, take place

When something *occurs*, it does so as an unscheduled event. When something *takes place*, it is a scheduled event.

EXAMPLES

    The snowstorm *occurred* very suddenly.
    The Cobb Corn Boil *takes place* every August.

of

Do not substitute *of* for *have*. Because *have* is unstressed in *could have, should have, would have, may have,* and other such constructions, it is often mistaken for *of.*

weak	improved
He *must of* been mistaken about the experiment.	He *must have* been mistaken about the experiment.

off of

Off of is the use of two prepositions when only the first one is necessary.

weak	improved
The worker fell *off* of the scaffolding.	The worker fell *off* the scaffolding.

OK, O.K., okay

Although all three are acceptable spellings, do not use the word in technical writing because it is conversational.

on account of

Do not use the conversational *on account of* as a substitute for *because of* in technical writing.

weak	improved
We didn't go to the fair *on account of* the rain.	We didn't go to the fair because of the rain.

on the grounds that, of

Both *on the grounds that* and *on the grounds of* are wordy, unnecessary substitutes for *because*.

weak	improved
He said he couldn't go *on the grounds that* he had too much homework.	He said he couldn't go *because* he had too much homework.

Note: *On the grounds that* is the correct phrase when you are explaining the basis of an argument.

onto, on to

On is a preposition that indicates a location. *Onto* is a preposition that indicates a movement to a position. *On to* does not mean the same thing that *onto* does.

EXAMPLES

He placed the ball *on* the deck.
He climbed *onto* the deck.
He went *on to* open the door for me.

orient, orientate

Both terms have the same meaning, but because *orientate* is the longer term, do not use it in technical writing. *Orient* is both timesaving and space-saving.

ourself
Ourself is not a word. *Ourselves* is the correct word.

outside of
Outside of is the use of two prepositions when only the first one is necessary.

weak	improved
He moved *outside of* the city limits.	He moved *outside* the city limits.

over with
Over with is the use of two prepositions when only the first one is necessary.

weak	improved
We wanted to get the presentation *over with*.	We wanted to get the presentation *over*.
	or
	We wanted to finish the presentation.

paid, payed
People often mistake *paid* for *payed*. *Paid* is the past participle for the verb *to pay*, which means *to give a person what is due for items received*. *Payed* is the past tense for the verb *to pay*, which means *to waterproof*.

parameter
Parameter is understood by many nontechnical people to be synonymous with *perimeter*—the distance around the outside of an object, such as the circumference of a circle or the boundaries of a piece of property. Because nontechnical audiences do not understand that *parameter* refers to *a quantity with variable values*, and because *parameter* has been overused in technical fields, avoid using the term in technical writing.

part, portion
A *part* is a fraction of the whole. A *portion* is an allotment.

EXAMPLES

He wondered what *part* he was going to play in the reorganization of his department.

He received his *portion* of the wager.

party
Use *party* to refer to a person only when using legal language, as in "*party* of the first part."

pardon   See **excuse, pardon.**

penultimate   See **ultimate, penultimate.**

per
Per is a business term that is used in two ways. First, it is used in Latin expressions such as *per capita* and *percent*. Second, it is used to mean *according to*, as in "As per your request." Here it becomes business-letter jargon that you should avoid.

percent, per cent, percentage
Percent has replaced the older spelling *per cent.* Use *percent* when written with numbers. Use *percentage* when numbers are not specified.

EXAMPLES

10 percent, *100* percent
a large *percentage*, a small *percentage* of the audience

personal, personnel
Personal is an adjective that refers to an individual. *Personnel* is a noun that refers to people who are employed in the same establishment. *Personnel* does not refer to people in general.

EXAMPLES

Hope resigned for *personal* reasons.
The *personnel* in the payroll department won the lottery.

persons, people
Persons are individual people. *People* are an anonymous group.

EXAMPLE

Only three *persons* stood out among the *people* who had assembled for the rally.

phenomenon, phenomena

Phenomenon is the singular form of *phenomena*. *Phenomena* must take a plural verb.

EXAMPLES

The *phenomenon* was observable from the platform. (one observable occurrence)
The *phenomena* were observable from the platform. (more than one observable occurrence)

plenty

Plenty is often used informally to mean *sufficient* or *abundant*. Avoid this use in technical writing.

weak	improved
The meal was *plenty* good.	The meal was delicious.

plus

Do not use *plus* as a substitute for *and* between main clauses, nor as a substitute for conjunctive adverbs.

weak	improved
Sandra and I went to the dinner *plus* we had a good time. (substitute for *and*)	Sandra and I went to the dinner *and* we had a good time.
Our department merged with marketing, *plus* we now have their department head. (substitute for conjunctive adverb)	Our department merged with marketing; moreover, we now have their department head.

PM, AM or pm, am   See **AM, PM or am, pm.**

poorly

Poorly is informal for *ill*. (Also see **badly.**)

weak	improved
I feel *poorly* today.	I feel *ill* today.

practicable, practical
Practicable indicates that something is possible or could be accomplished. *Practical* indicates that something is not only possible but useful as well.

EXAMPLES

It was a *practical* solution that solved many problems for the company.

The can crusher was an interesting invention, but the research and development department wondered if it was *practicable* because of its expense.

prejudice, bias   See **bias, prejudice.**

prep
Prep is informal usage or jargon for words such as *prepare* and *preparatory*. Do not use *prep* in technical writing; write out the full word instead.

presently, at present
Presently means *soon*. *At present* means *right now*. *Presently* should never be substituted for *at present*.

EXAMPLES

The plant will open *presently*. (soon)
At present, the plant is open. (right now)

principal, principle
Principal is both a noun and an adjective. When used as a noun it refers to the head of a school or to a sum of money. When used as an adjective it refers to first in rank. *Principle* is always a noun that refers to a fundamental law or a basic truth.

EXAMPLES

The *principal* witness disappeared.
The *principal* of the high school was fired.

I received 6 percent interest on the *principal* in my account at First Federal.
She believed in the *principle of equality.*

pseudo, quasi, semi

Pseudo is a prefix meaning *counterfeit. Quasi* is a prefix meaning *partial. Semi* is a prefix meaning *half or twice. Quasi* is often confused with *pseudo* and *semi.* Neither *pseudo* nor *semi* need a hyphen; *quasi* needs a hyphen when it serves as a prefix to an adjective.

EXAMPLES

pseudonym, pseudointellectual, pseudoscience
quasi-judicial, quasi-intellectual, quasi science
semicircle, semiformal, semiliterate

quote

Quote is informal for *quotation.* In technical writing, use *quotation.*

EXAMPLES

direct *quotation, quotation* marks

rarely ever, hardly ever

In technical writing use one of two substitutes for *rarely ever* or *hardly ever: rarely* or *hardly.*

weak	improved
He *rarely ever* attended the opera.	He *rarely* attended the opera.
She *hardly ever* noticed me.	She *hardly* noticed me.

re

Re or *re* is jargon for *regarding* and was once frequently employed in both letter and memo writing. The term is disappearing from use and has been replaced by *subject.* Both terms are used in headings on memos.

EXAMPLES

To: Hope Goodheart
From: Edgar Allen
Re: Carlson contract

To: Hope Goodheart
From: Edgar Allen
Subject: Carlson contract

real
Real, when used to mean *very,* is a conversational usage that you should avoid in technical writing.

reason is because
This usage is redundant for *because.* Avoid *reason is because* in technical writing.

weak	improved
We haven't finished our report, but the *reason is because* we have been too busy.	We haven't finished our report *because* we have been too busy.

reckon
Reckon is an informal usage for *think.*

weak	improved
I *reckon* the office will close on Veterans Day.	I think the office will close on Veterans Day.

refer   See **allude, refer.**

regardless, irregardless   See **irregardless, regardless.**

relate to
Relate to is an overused informal expression that means *to be sympathetic to.* Avoid *relate to* in this sense in technical writing.

remainder   See **balance, remainder.**

respectively, respectfully
Respectively refers to the order in which things are designated. *Respectfully* is a term of respect.

EXAMPLES

The president, vice-president, secretary, and treasurer are Betty, Norma, Joanne, and Cora, *respectively.*
We all treat our vice-president *respectfully.*

scared, afraid, frightened   See **frightened, scared, afraid.**

seldom ever

Seldom ever is a redundancy for *seldom* . Avoid *seldom ever* in technical writing.

semi   See **bi, semi** and **pseudo, quasi, semi.**

shall, will

Shall is an extremely formal usage that expresses a command. *Will* is an indication of the future that is used in place of *shall,* even in formal writing. Most writers use *will.*

EXAMPLES

> I *shall* return.
> I *will* return

sic

Sic is Latin for *thus.* It is used to indicate that a writer has reported something exactly as it was incorrectly written in an original statement or document, such as a misspelled word or an inaccurate number. Place *sic* in brackets after the incorrect word or number.

EXAMPLES

> Jeremy wrote on his test that Hemmingway [*sic*] is his favorite author.
> Dr. Anderson wrote that the moon was 2000 [*sic*] miles from Earth.

site, sight, cite   See **cite, site, sight.**

-size, -sized

Both *-size* and *-sized* are suffixes that are not needed in most writing because they are redundant. An *oversized* chair is not redundant, but *large-size* chair is, for *large* chair is more specific and shorter.

so

So is a vague word that you should not use if another more exact word can be substituted for it.

weak	improved
I wasn't promoted, so I am looking for another job.	*Because* I wasn't promoted, I am looking for another job.

so, so that

In technical writing *so that* is the preferred usage because *so* can be ambiguous.

weak	improved
We stayed home so we could watch the World Series.	We stayed home so *that* we could watch the World Series.

In the *weak* example, it isn't clear if the subjects stayed home on purpose to watch the World Series or if *because* the subjects stayed home, they watched the World Series. In the *improved* example, *so that* clarifies that the subjects stayed home intentionally.

some, somewhat

Some is an adjective that means *unspecified. Somewhat* is an adverb meaning *a little.* Do not confuse the two terms.

EXAMPLES

Some of us are in agreement about what to do with the Ferguson account.

We are *somewhat* divided about what to do with the Ferguson account.

someone, some one   See **anyone, any one, everyone, every one for usage explanation.**

sort of   See **kind of, sort of.**

specially, especially   See **especially, specially.**

stationary, stationery

Stationary refers to something that is in a fixed position. *Stationery* refers to paper and envelopes for writing letters.

EXAMPLES

Once the central computing system was installed, it was obvious it would be a *stationary* part of the company's equipment.

Helen borrowed some *stationery* to write a letter.

subsequently, consequently
Subsequently means that something happens later or after something else. *Consequently* means that something happens as a result of something else.

EXAMPLES

> We missed the meeting last Friday; *subsequently*, we learned that we didn't hear the new proposal.
> We missed the meeting last Friday; *consequently*, we didn't hear the new proposal.

supplement   See **augment, supplement.**

suppose to, supposed to
Some writers make the mistake of not adding a *-d* to *suppose*. The correct use is *supposed to*.

weak	**improved**
I was *suppose to* attend the symposium.	I was *supposed to* attend the symposium.

sure
Do not use *sure* in place of *surely* or *certainly*.

weak	**improved**
We will *sure* go bowling with you on Saturday night.	We will *surely* (or *certainly*) go bowling with you on Saturday night.

tacit, explicit, implicit   See **explicit, implicit, tacit.**

take place, occur   See **occur, take place.**

tenant, tenet
A *tenant* is a person who rents or occupies the property of another person. A *tenet* is a belief held by a person or a group.

EXAMPLES

The *tenant* moved when his neighbors complained about his drum-playing at 6 am every morning.

One of the group's *tenets* was that work was harmful.

than, then

Than and *then* are frequently confused. Use *than* in comparisons. Use *then* to indicate time.

EXAMPLES

My boss is easier to work for *than* the boss in marketing.

Even *then* it was too late to do anything.

that, which, who

That is an overused term for both *which* and *who*. You usually can distinguish the three terms as follows: Use *who* with people; use *that* and *which* with animals, objects, and ideas. In addition, use *that* in restrictive clauses (clauses that are necessary for the sentence). Use *which* in nonrestrictive clauses (clauses that add material to a sentence but are not necessary to the sentence).

EXAMPLES

restrictive clauses

He is the man *who* owns the local Dairy Queen.

Avocados are one food *that* Sally won't eat.

nonrestrictive clauses

Veronica, *who* has been my secretary for five years, announced that she was moving to Hawaii next month.

The fire, *which* the investigators ruled as arson, destroyed three old buildings on K Street.

their, there, they're

Their is a possessive pronoun; *there* is a location; *they're* is a contraction for *they are*. These sound-alike words are frequently misspelled or misused.

EXAMPLES

Their new house is on Elm Street.
There is the man who sold me my new car.
They're vacationing in Florida this winter.

theirself, theirselves
Neither *theirself* nor theirselves is a word. *Themselves* is the correct term.

then   See **than, then.**

these kind
These is plural, while *kind* is singular. Do not use the two words together.

this here, that there, these here, them there
All of the above expressions are nonstandard. Eliminate *here* from each of them.

thusly
Thus is already an adverb, so it is redundant to add an *-ly* to it. Use *thus*, not *thusly*, in technical writing.

till, until
In technical writing *till* is not an acceptable substitute for *until.*

to, too, two
These three sound-alike words are constantly confused by writers. *To* is usually a preposition; *too* means *also*, and *two* is a number.

EXAMPLES

Give the book *to* me.
He had *too* many drinks at the office party.
Two of us will attend the committee meeting.

tortuous, torturous
Tortuous means *twisting or winding, as a road. Torturous* means *painful* or *agonizing.*

EXAMPLES

The *tortuous* road caused us to be fifteen minutes late to an appointment.

After a *torturous* game of rugby, Curt was in great pain.

toward, towards
Both words are correct, although Americans tend to prefer *toward,* while the British prefer *towards.*

try and
Try and is not a logical construction any more than *I want and.* The expression should be avoided in all writing.

weak	improved
Try and stop me.	Try to stop me.

type
Type is a noun that you should not use as an adjective unless it is hyphenated.

weak	improved
We decided to try the new *type* directory of corporations.	We decided to try the new *type* of directory of corporations.
She has blood *type* O.	She has blood-*type* O.

ultimate, penultimate
Ultimate means *last*, not best. *Penultimate* means *second to the last*, not the last.

uninterested, disinterested   See **disinterested, uninterested.**
unique
Unique is an often misused term that means it is one of a kind. Writers often add superlatives to the term. If something is already one of a kind, it cannot be the *most unique* or *more unique. Unique* always should be used by itself.

weak	improved
Sue has the *most unique* way of insulting people.	Sue has a *unique* way of insulting people.

utilize

Do not substitute *utilize* for *use.*

weak	improved
She *utilized* the new copy machine for an hour.	She used the new copy machine for an hour.

very

Very is often unnecessary. It is an *intensifier*—that is, it is a word that defines or emphasizes the degree of another word. Too many intensifiers can weaken the specificity of a word and should be avoided. Eliminate *very* and substitute the two words with a stronger word wherever possible.

weak	improved
Ty was *very angry* when he learned that Arne had been promoted to district manager.	Ty was furious when he learned that Arne had been promoted to district manager.

wait on

Wait on is colloquial usage for *wait for.*

weak	improved
We are *waiting on* Jim's arrival.	We are waiting for Jim's arrival.

want in, want out

Want in and *want out* are conversational for *want to enter* and *want to leave.* Do not use either one in technical writing.

weak	improved
Larry *wants in* on our group.	Larry wants to enter our group.
Larry *wants out* of our group.	Larry wants to leave our group.

ways

Ways is conversational for *way.* Do not use *ways* in technical writing.

weak	improved
It is a long *ways* home.	It is a long way home.

weather, whether
Because they sound alike, writers often confuse these two terms. *Weather* refers to atmospheric conditions. *Whether* is used to set up alternatives.

EXAMPLES

The *weather* has been unusually mild for this time of year.
She couldn't decide *whether* she should major in computer science or mathematics.

well See **good, well.**

where
Where is a conversational substitute for *that.* Avoid this usage in technical writing.

weak	improved
I noticed where Pat is now wearing bow ties.	I noticed that Pat is now wearing bow ties.

where . . . at, where . . . to
Do not use *at* or *to* with *where,* as in "Where's it at?" Avoid using either of these prepositions with *where* because they are not needed.

which See **that, which, who.**

will, shall See **shall, will.**

-wise
-Wise is an overused suffix that you should avoid in technical writing because of the generally unacceptable terms that are formed with this suffix.

EXAMPLES

weatherwise, electronicwise, computerwise

In a few cases, the use of *-wise* is a necessity, as in *clockwise* and *counterclockwise.*

with regards to See **in regards to, with regards to.**

would of
Would of is nonstandard for *would have*. For a more complete discussion, see **of**.

your, you're
Your is the possessive of *you*. *You're* is a contraction for *you are*.

EXAMPLES

Put *your* coat in the closet.
Norman told me *you're* going to leave in two weeks.

INDEX

A 0
B 1
C 2
D 3
E 4
F 5
G 6
H 7
I 8
J 9

(continued from inside front cover)